RAND NATIONAL DEFENSE RESEARCH INSTITUTE

T0210405

Designing Adaptable Ships

Modularity and Flexibility in
Future Ship Designs

John F. Schank, Scott Savitz, Ken Munson, Brian Perkinson,
James McGee, Jerry M. Sollinger

Prepared for the United States Navy

Approved for public release; distribution unlimited

For more information on this publication, visit www.rand.org/t/RR696

Library of Congress Cataloging-in-Publication Data is available for this publication.
ISBN: 978-0-8330-8722-5

Limited Print and Electronic Distribution Rights

Support RAND

Make a tax-deductible charitable contribution at
www.rand.org/giving/contribute

www.rand.org

Preface

The U.S. Navy faces challenges in acquiring and supporting the numbers and types of ships needed to meet national security requirements. Ships are expensive, often costing billions of dollars, and the Navy faces tough choices each year in determining what ships to procure with limited defense funds. These challenges and choices would be difficult enough in a static world. However, adversaries, missions, and technologies change, requiring new capabilities for naval warships. Because of the high cost of acquiring new warships, the Navy seeks to extend the service lives of operational ships by adapting them to cope with new adversaries, carry out new missions, and incorporate new technologies. But such modernizations are also expensive and may be constrained by the available support systems and/or configuration of the ship.

The concepts of modularity and flexibility are championed as ways to potentially reduce both the time and cost of modernizing in-service ships and to adapt to future uncertainties. The Navy has used these concepts in a number of programs, and they are now more broadly discussed and more widely considered in ship design and construction programs. But modularity and flexibility imply different things to different people and may not—in and of themselves—provide future adaptability. Also, there are hurdles to a wider consideration of modularity and flexibility concepts across ship classes.

To gain a better understanding of the concepts of modularity and flexibility in ship design, the Program Executive Officer for Ships tasked the RAND Corporation to develop a path that would more firmly ingrain those concepts into future ship design and construction programs, with the overall objective of reducing the costs of modernizing ships to meet new missions or to accommodate new technologies.

This research was sponsored by the U.S. Navy and conducted within the Acquisition and Technology Policy Center of the RAND National Defense Research Institute, a federally funded research and development center sponsored by the Office of the Secretary of Defense, the Joint Staff, the Unified Combatant Commands, the Navy, the Marine Corps, the defense agencies, and the defense Intelligence Community. For more information on RAND's Acquisition and Technology Policy Center, see http://www.rand.org/nsrd/about/atp.html or contact the director (contact information is provided on the web page).

Contents

Figures and Tables

Figures

Tables

Summary

The U.S. Navy faces a number of challenges as it decides how many and which types of warships to buy. Ships are expensive, with some of the larger ones costing billions of dollars. Simultaneously, funds are limited and likely to become more so as budgets shrink in the aftermath of the Iraq and Afghanistan conflicts. Furthermore, ships remain in the fleet for a long time, often three decades or longer. It is difficult for the Navy to anticipate exactly what missions its fleet will need to carry out that far into the future. As missions and technologies change, the typical response is to modernize ships to accommodate the new mission or technology. But modernization is expensive, and the physical configuration of the ship may limit what can be done. In some cases, the modernization costs are so high that the Navy has determined it is cheaper to retire the ship than to modernize it.

The concepts of modularity and flexibility are discussed as ways to design and build adaptable ships that will reduce modernization costs, and perhaps as ways to reduce initial costs. *Modularity* entails partitioning a system into modules that consist of self-contained elements. It hinges on a systems engineering process that stresses functional analysis and identification of key interfaces. Typically, the concept calls for using common industry standards for key interfaces. *Flexibility* is a broader, less-precisely defined concept, but generally means constructing ships in such a way that they can more readily adapt to changing missions and technologies. Modularity can be a subset of flexibility and together they contribute to adaptable ships.

Our study of modularity concepts does *not* focus on construction modularity in shipbuilding. All shipbuilders currently use large construction pieces (often called modules, blocks, or rings) when building their ships. Construction modularity aims to reduce the cost of building ships and is not necessarily aimed at the ease of upgrading ships due to new technologies or missions.

Focus of This Research

While the Navy has adopted modularity and flexibility concepts in several programs, it has not implemented these concepts in ship design universally. In part, this has occurred because no consensus exists in the Navy on how these concepts should be collectively applied. Furthermore, since modularity and flexibility normally imply larger ships, the current cost-estimating practices may suggest increased acquisition cost. Yet, life cycle costs could be less if a larger ship's design was more adaptable, despite being larger in size and weight. To determine future directions for designing adaptable ships, Navy decisionmakers need to answer the following three questions:

- What modularity and flexibility concepts should the Navy apply in future shipbuilding programs?
- What changes in a ship's capability may be needed in the future?
- When are the future opportunities to adapt modular and flexible designs in future naval ship programs?

To address these questions, the Program Executive Officer for Ships tasked the RAND Corporation to develop a roadmap that would more firmly embed the concepts of modularity and flexibility into future ship design and construction programs. Specific research issues include the following:

- Develop a method to categorize various modularity and flexibility options
- Estimate the costs of the lack of modularity and flexibility in ship design
- Understand ways to achieve modularity and flexibility when facing an uncertain future

- Project when in the future opportunities exist to adopt the concepts of modularity and flexibility.

Findings

Modularity and Flexibility Concepts

The concepts of modularity and flexibility have different meanings and interpretations. Modularity involves creating fixed boundaries, defined interfaces, and defined ship services (such as power and cooling) to standard portions of a ship, which are termed *modules*. We define three types of modularity:

- **Common modules used across multiple classes of ships.** These common modules are structural pieces of the ship that are built and tested in a factory-like environment. Although not currently adopted in Navy ship designs, potential applications include hotel-like functions such as galleys, medical facilities, and laundries.
- **Self-contained modules that provide a plug-and-play capability for the equipment inside the module.** These modules have defined interfaces and boundaries and are designed for a specific task, such as firing a missile. Where common modules can be used across different classes of ships, self-contained modules are typically used within a single class of ships. The vertical launch system (VLS) modules on *Arleigh Burke*–class destroyers are an example of a self-contained module.
- **Modular installations that provide a basic ship structure and services that allow various mission packages to be installed and interchanged as needed.** Modular installations, like self-contained modules, have defined interfaces but much broader defined boundaries. The U.S. Navy LCS and the Royal Danish Navy's *Absalon* class ships are examples of this type of modularity.

Flexibility involves the ability to change boundaries, whether they are physical or related to ship services. We define three types of flexibility:

- **Flexible infrastructures that allow changes to the boundaries of ship spaces to be made more quickly.** Flexible infrastructures use standard tracks, panels, and interfaces to allow the movement of bulkheads and the connection of ship services such that spaces can be reconfigured to meet evolving needs. This type of flexibility is used on *Ford*-class aircraft carriers.
- **Additional space within a ship.** Where flexible infrastructures allow the interior boundaries of a space to be adjusted, expanding the space within the ship can also provide future flexibility. This type of flexibility implies larger ships.
- **Additional ship services within a space.** Modernizations for new missions or technologies typically involve adding power, cooling, and fiber-optic hookups to the equipment within a space. Having extra power, cooling, and fiber-optic drops within a space increases the flexibility of the ship to address future modernizations at a lower cost.

The Navy has begun to incorporate adaptability concepts into ship designs, and senior leaders suggest modularity and flexibility are needed in naval ships. The Navy needs to continue this trend and advance the thinking on cost-effective, adaptable ships.

Cost Effects of Modularity and Flexibility

There are different points of view on the effect of modularity and flexibility concepts on the design and construction of a naval ship, with little technical analysis to support those views. On the one hand, design costs may be greater for the initial development of a common or self-contained module, but use of the module on subsequent classes of ships may reduce design and construction costs in the future. Construction costs can be reduced through the economies of scale provided by common components and by any reduction in welding and other hot work done in a factory versus during ship construction. On the other hand, flexibility that leads to larger ships could also lead to higher acquisition costs.

Modularity and flexibility could have their biggest influence on the cost of modernizing a ship during her operational life. Modularity and flexibility should have little or no effect on the equipment cost of a modernization work package, but can have a large effect on the cost of installing and testing the new equipment or system. Installation person hours are driven by the amount of welding and other hot work needed to remove and replace the equipment because of structural restrictions and various interferences. Future designs should consider the accessibility (welded access plates, interference removal, etc.) of any equipment that has a high likelihood of being replaced during the ship's life. Common foundations and standard connections and arrangements for various ship services, such as power, cooling, and data transfer, should be developed and used on future ship designs.

Although not related to modularity or flexibility, better coordination and management of repair and modernization work packages could lead to lower costs because of the elimination or reduction in duplicative activities. We understand the Navy is moving toward a single program manager for repair and modernization work packages.

How Ships May Have to Adapt in the Future

Geopolitically, there is a great deal of uncertainty over any multi-decade horizon. It is important that ships have the ability to contribute in very different types of conflicts against adversaries ranging from near-peers to non-state actors, some of which may not be perceived as possible threats when the ship is built. For example, ships built just a quarter-century ago, in the late 1980s, were designed with the expectation that the Cold War against the Soviet Union would continue; moreover, given the legacy of the Vietnam War, the nation appeared likely to avoid involving itself in counterinsurgencies and civil wars, if at all possible. However, U.S. Navy ships have been able to contribute effectively in Iraq, the Balkans, and Afghanistan, despite the land-centric nature of these conflicts. They have also helped to deter adventurism by regional and rising powers.

The coming decades may be as unpredictable as, or even more unpredictable than, the past few. The U.S. military may face a wider range of potential adversaries, as some states may rise to near-peer levels and non-state actors acquire more advanced capabilities. U.S. Navy ships need to be prepared to accommodate the diverse mission sets needed to counter such a range of adversaries.

Geopolitical uncertainty strengthens the case for increasing margins of available ship services—power, cooling, support for personnel, space, and bandwidth. By having ample reserves of each of these, ships can accommodate changing missions, as well as novel technologies or more advanced versions of existing ones. The case for "bigger is better" needs to be considered more widely. Additional space will provide room to add power, cooling, and personnel, as well as new mission-specific systems. A key challenge will be to understand where the additional space will be needed and how best to use available space until it is needed, designing internal boundaries accordingly.

We assessed that four major technological trends will likely influence naval operations over the coming decades:

- The rapidly increasing use and effectiveness of off-board unmanned systems
- The growing importance of using of the electromagnetic spectrum as a weapon
- Enhanced capabilities for long-range targeting
- The increasingly networked nature of the battlespace.

All of these trends involve rapid change and inherently unforeseeable technological developments. However, by endowing ships with more capacity or capability than immediately needed in five areas, it is possible to make them better able to accommodate these trends. These five areas are **power, cooling, support for personnel, space, and bandwidth.** To varying degrees, greater margins in these attributes can also be mutually supporting. For example, additional space can be used to support extra personnel or to amplify power output; the availability of more power and personnel, in turn, can enable the effective use of greater quantities of bandwidth. Of course, new designs must include sufficient weight and stability margins to accommodate future ship modifications.

Future Opportunities for Modularity and Flexibility

The primary place to inject modularity and flexibility concepts is early in the ship concept design phase. Adding these concepts to existing designs is difficult. This is one reason for the Navy's decision to retire cruisers rather than modify them. In the past, the Navy typically had a number of designs underway, with one design following close upon another, but this is no longer true. Ships remain in the fleet longer, and budgets are shrinking. These influences contribute to gaps in current and future ship design. In the near term, only two opportunities for new or updated designs present themselves: the DDG-51 Flight III and the LX(R) programs. New designs beyond those programs are a decade or more away, but efforts should begin on how to design adaptability into the future class of surface combatants. For example, the Navy has started a new frigate program based on the LCS class. This new ship class should consider how modularity and flexibility can be integrated into the new design while reducing the construction cost of the ship.

Recommendations

We offer both short-term, ship-specific recommendations and long-term, more general recommendations.

Short-Term, Ship-Specific Steps
DDG-51 Flight III
This program offers fewer opportunities to accommodate modularity and flexibility innovations than the LX(R). However, limited opportunities do exist. First, as interior spaces are changed, new walls and floors could adopt the same flexible infrastructure track and interface concepts used in the *Ford*-class carrier design. Second, the Navy could design and build common hotel-related modules that could ultimately be used across multiple ship classes.

LX(R)
This ship class offers more opportunities for incorporating modularity and flexibility than the DDG-51. The length of LSDs has grown

from the *Thomaston* class (510 feet) to the *Whidbey Island* and *Harpers Ferry* classes (610 feet). Flexibility suggests the LX(R) should be even longer and have more power-generation capability. The *San Antonio* class of LPD ships, with a length of 684 feet, has been examined as an alternative for the LX(R) and warrants a detailed evaluation. The LX(R) design should continue to incorporate various payload modules. It should also consider large, open hangar bays with connections for those payload modules. Finally, it should consider flexible infrastructures and common modules that can be used across multiple ship classes.

Overarching Recommendations

We recommend that the Navy do the following:

- Continue to encourage and develop the concepts of modularity and flexibility but in a more focused and coordinated fashion. Evaluate the potential for common foundations for certain classes of equipment (such as the standard racks used for computer-related equipment) and standard connections for the ship service interfaces between spaces. Also, evaluate the potential use of common hotel-related or other types of modules for use across different classes of ships. For example, a scalable modular concept within the design production model tool for a future class of ships could be exported for use across other ship designs. The use of a rail gun is an example of a module on future ship classes.

- Continue to develop a single organization to manage repair and modernization work packages in a coordinated manner. This organization should also have the mission of developing future modularity concepts that cut across ship classes and initiate conceptual and feasibility designs for interfaces and payload modules. It should work with the major weapon system developers and manufacturers.

- Collect, organize, and analyze modernization work package data to gain a better understanding of the cost of modernizing in-service ships, including what drives costs and how those costs could be reduced if more modularity and flexibility were incorporated into ship designs.

Acknowledgments

We thank RADM Dave Lewis, Program Executive Officer for Ships, for his continued support and interaction throughout this report. Dr. Norbert Doerry, technical director of the Naval Sea Systems Command's (NAVSEA's) Engineering Directorate's Technology Group, and Patrick Karvar, Naval Surface Warfare Center, Carderock Division (NSWCCD), provided background information and shared their knowledge and expertise during the course of the research. Jack Abbott of AOC Incorporated provided a historical perspective of modularity in the U.S. Navy and shared his wealth of experiences in adopting modularity concepts. CAPT Mark Vandroff, DDG-51 Program Manager, facilitated our early interacts with the DDG-51 program. Special thanks goes to CDR Sam Pennington, PMS 400F, for providing outstanding support in data collection and understanding of the process.

We would also like to thank Huntington Ingalls Industries–Newport News Shipbuilding for describing the flexible infrastructure system, and BAE Systems for providing information on the modernization installation process as well as insights on what drives installation costs. We thank Bath Iron Works for helping obtain and understand the mid-life modernization data for the USS *John Paul Jones*. Many others, too numerous to mention, shared their time and expertise with us.

We appreciate the helpful comments provided by Robert Murphy and Michael McMahon on an early draft of the report. Their suggestions and comments helped strengthen the overall report.

Of course, any errors of omission or commission are the responsibility of the authors alone.

Abbreviations

ACB12	Advanced Capability Build 12
ACS	Aegis combat system
AIMS	Architectures, Interfaces, and Modular Systems
AIT	Alteration Installation Team
ASW	anti-submarine warfare
ATC	Affordability Through Commonality
AWS	Aegis weapon system
BL	baseline
BMD	ballistic missile defense
C4ISR	command, control, communications, computing, intelligence, surveillance, and reconnaissance
CANES	Consolidated Afloat Networks and Enterprise Services
CBRN	chemical, biological, radiological, and nuclear
CIC	combat information center
CG	guided missile cruiser
CNO	Chief of Naval Operations
COATS	Command and Control System Module Off-hull Assembly and Test Site

COTS	commercial off the shelf
cpm	cubic feet per minute
DDG	guided missile destroyer
DoD	Department of Defense
ECP	engineering change proposal
GFM	government-furnished materiel
gpm	gallons per minute
HM&E	hull, mechanical, and electrical
HVAC	heating, ventilation, and air conditioning
ISR	intelligence, surveillance, and reconnaissance
IT	information technology
KG	measure of ship's vertical center of gravity
LCS	littoral combat ship
LPD	landing platform dock or amphibious transport dock
LSD	dock landing ship
MAS	modular adaptable ship
MCM	mine countermeasures
MEKO	Mehrzweck-Kombination
MMSP	Multi-Mission Signal Processor
MOSA	Modular Open Systems Approach
MSSIT	Mission Systems and Ship Integration Team
NAVSEA	Naval Sea Systems Command
NSWCCD	Naval Surface Warfare Center Carderock Division

OACE	Open Architecture Computing Environment
OSA	Open Systems Architecture
OSJTF	Open Systems Joint Task Force
scfm	standard cubic feet per minute
SEAMOD	Sea Systems Modification and Modernization by Modularity
SSES	Ship Systems Engineering Standards
SUW	surface warfare
TI12	Technology Insertion 12
TOSA	Total Ship Open Systems Architecture
UAS	unmanned aircraft system
USV	unmanned surface vehicle
UUV	unmanned undersea vessel
VLS	vertical launch system
VPS	variable payload ship

CHAPTER ONE

Introduction

The Navy's Dilemma

The U.S. Navy is being asked to do more with less. Budget constraints pose challenges in managing the shipbuilding program in a way that provides the number and types of ships and submarines needed to meet current and future national security objectives while sustaining a shipbuilding industrial base. Naval ships are expensive, often costing billions of dollars to buy and operate. Somewhat offsetting the high initial acquisition cost is the fact that naval ships are designed and built to remain in service for multiple decades. The majority of ships in the current fleet were built over 20 years ago, and many are far older. For example, the USS *Enterprise*, commissioned in the 1960s, was inactivated only in 2012 after serving the fleet for 50 years.

Theoretically, the long operational lives of naval ships help the Navy meet force-level objectives because few new ships are needed each year. However, these long operational lives pose their own set of challenges for the Navy. The ships in the operating fleet must adapt to new missions, threats, and technologies. Building a ship takes several years, and many newly constructed ships—especially first-of-class ships—must almost immediately undergo system upgrades to accommodate new and evolving technologies. Ship classes with long production runs, such as *Los Angeles* and *Virginia* class attack submarines and *Arleigh Burke*–class destroyers, are typically built in what are called flights, with each successive flight involving several major upgrades and changes. For example, the Navy's DDG-51 program is currently going through the design phases for Flight III, more than two decades after the lead ship was delivered to the Navy.

Even ships within the same flight require updates to their equipment and systems. Ships undergo several upgrades during their operational lives, including a major mid-life modernization process to incorporate new technologies, systems, and equipment. These modernization efforts typically have high costs, so much so that at times the Navy may determine that it is more fiscally prudent to retire a ship earlier than planned than it is to spend the money to upgrade the ship.[1] The challenge the Navy faces is to design and build ships in a way that reduces the time and cost required to incorporate changes due to new missions and technologies.

Modularity and Flexibility

The concepts of modularity and flexibility are often championed as solutions to the challenges of high construction and modernization costs for both new and in-service ships. They are related but distinct concepts. As defined by a Naval Sea Systems Command (NAVSEA) document, modularity is a design approach in which a system has the following characteristics:

- functionally partitioned into discrete, scalable, and reusable modules consisting of isolated, self-contained elements
- a systems engineering process that emphasizes functional analysis and the identification of key interfaces
- common industry standards for key interfaces to the largest extent possible.[2]

---

[1] The Navy has proposed retiring five guided-missile cruisers rather than incur the heavy cost of maintaining their missile launching systems. These initial *Ticonderoga*-class cruisers were equipped with twin-armed missile launchers. Later ships in the class had the improved and modular Mark 41 vertical launch system. See Megan Eckstein, "Shannon: Navy Ready to Upgrade Cruisers if Congress Provides Funding," *Inside the Navy*, May 12, 2012.

[2] See Patrick Karvar, Shawna Garver, Ray Marcantonio, and Philip Sims, *Modular Adaptable Ship (MAS) Total Ship Design Guide for Surface Combatants*, Washington, D.C.: Naval

Flexibility is not specifically defined in authoritative naval documents, but implies mission adaptability through future modernizations and upgrades.

Modularity and flexibility are not new concepts to the Navy and have been adopted in several programs. *Arleigh Burke*–class destroyers (DDG-51) were designed with a vertical launch system (VLS) that can adapt to various missile systems. The *Ohio*-class ballistic missile submarines were designed with bigger missile tubes than needed for the C4 missile to allow easier conversion to the larger D5 missile. The *Spruance*-class destroyers were designed to be bigger than needed and were the basis for *Ticonderoga*-class cruisers. And, currently, the LCS is designed as a basic sea frame with the space, services, and interfaces to accommodate several different mission module packages.

The Navy understands the importance of modularity concepts in the design and construction of future ships. ADM Jonathan Greenert, Chief of Naval Operations (CNO), in an article for the U.S. Naval Institute's *Proceedings* magazine, described the need to move from expensive, "luxury car" ships to "trucks" that could change mission payloads based on operational needs.[3] He also described how "substantial volume, reserve electrical power, and a small number of integral systems" are important to adapt to future requirements. Other Navy leaders have spoken about the need for modularity in the future naval fleet.[4] These concepts of "bigger" and "more," in conjunction with modularity, may provide a truly adaptable naval platform.

Research Focus and Report Organization

The goal of greater modularity and flexibility has influenced, but not necessarily governed, ship design and construction. The paradigms have been implemented on a program-by-program basis rather than as

---

Sea Systems Command, February 2011.

[3] See Jonathan W. Greenert, U.S. Navy, "Payloads over Platforms: Charting a New Course," *Proceedings*, Vol. 138/7/1,313, July 2012.

[4] See Eckstein, 2012.

a universal ship design philosophy.[5] There are arguments on both sides of the issue. Most proponents suggest lower acquisition costs in addition to less expensive life cycle costs. Since modularity and flexibility typically imply larger and heavier ships, current acquisition cost estimation models, which are primarily weight-based, suggest higher construction costs. Additionally, reductions in the cost of major mid-life modernizations because of increased modularity and flexibility have proven difficult to substantiate, since there have been no historical cost analyses. In light of these uncertainties, program managers often have difficulty justifying increased acquisition costs to offset potentially lower, though unknown, future modernization costs.

In this period of strong interest in modularity and flexibility in future naval ships, decisionmakers must answer three questions:

- What modularity and flexibility concepts should the Navy apply in future shipbuilding programs?
- What changes in a ship's capability may be needed in the future?
- When will there be opportunities to adapt modular and flexible designs in future naval ship programs?

To answer these questions, the Program Executive Officer for Ships tasked the RAND Corporation to help more firmly ingrain those concepts into future ship design and construction programs. Our study of modularity concepts does not focus on construction modularity in shipbuilding. All shipbuilders currently use large construction pieces (often called modules, blocks, or rings) when building their ships. Construction modularity aims to reduce the cost of building ships and is not necessarily aimed at the ease of upgrading ships due to new technologies or missions. Specific research issues include the following:

---

[5] Modularity concepts have gained the greatest foothold in the information technology area. The Navy recently published an open-architecture strategy outlining a plan to improve the commonality across ship programs. See Geoff Fein, "USN Issues Open System Strategy to Foster Component Use Across Multiple Platforms," *International Defence Review*, January 11, 2013.

- Categorize various modularity and flexibility options.
- Estimate the costs of the lack of modularity and flexibility in ship design.
- Understand ways to achieve adaptability when facing an uncertain future.
- Project when future opportunities will exist to adopt the concepts of modularity and flexibility.

This report describes the findings and recommendations of the research. Chapter Two describes the concepts of modularity and flexibility, how those concepts have been applied in different programs, and the current status of the Navy's modularity and flexibility efforts. Chapter Three describes the cost issues surrounding modularity and flexibility and provides the result of an initial data analysis. Chapter Four examines current technological and geopolitical trends, as well as historical lessons, in shaping modularity and flexibility programs for future ships. Chapter Five provides key findings and recommendations from the research, including describing a roadmap for incorporating concepts that lead to more modular and flexible naval ships. Four appendixes provide details on selected topics. Appendix A describes several modularity programs undertaken in the recent decades. Appendix B examines *Arleigh Burke*–class destroyers and how modularity was incorporated into the design of the ships. Appendix C describes the flexible infrastructure system. Appendix D describes the various work packages in the mid-life modernization of USS *John Paul Jones* (DDG-53).

CHAPTER TWO

Understanding the Concepts of Modularity and Flexibility

Many studies and reports over the past few decades have espoused the concepts of modularity and flexibility in naval ship design.[1] A common conclusion in all of these studies is that modularity and flexibility will more quickly embed new missions and technologies into the fleet while decreasing the total ownership cost of a ship, especially the cost of modernization to incorporate new technologies. The Navy has reacted to such studies by including modularity (but not necessarily flexibility) concepts in several programs. This chapter briefly describes several previous and ongoing programs that focused on modularity or open systems architectures. The chapter then presents a categorization of modularity and flexibility and describes previous and ongoing design programs that have adopted these various forms of adaptability. The chapter concludes with an assessment of the current status of modularity and flexibility applications in ship design.

Navy Programs with a Focus on Modularity

The concept of modularity has been the subject of several Navy and Office of the Secretary of Defense efforts over the past three decades.

---

[1] There are several strong proponents of modularity in U.S. Navy circles, including Jack Abbott, David Singer, Andrew Levine, and Norbert Doerry. A number of their articles and papers are provided in the report's list of references.

Figure 2.1 shows these various modularity programs.[2] NAVSEA initiated the Sea Systems Modification and Modernization by Modularity (SEAMOD) program in 1972. It was the first Navy study to examine a modular shipbuilding concept. SEAMOD examined the concept of designing a ship to receive a modular combat system to lower life cycle costs, introduce ships and weapon systems faster, and maximize fleet effectiveness. The program suggested that the advantages of modularity outweighed any disadvantages and that a modular ship would be more effective than a ship that did not incorporate modularity concepts. SEAMOD led to continuing studies on modularity in the Ship Systems Engineering Standards (SSES) program.

Figure 2.1
Modularity-Related Navy Programs

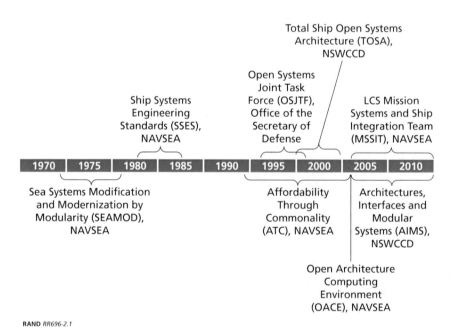

RAND RR696-2.1

---

[2] Appendix A provides more details on these various programs. In addition to U.S. Navy programs that focused on modularity, several foreign shipbuilders and navies have also adopted modularity principles. These include Germany's MEKO ships and the STANFLEX ships of the Royal Danish Navy.

The SSES program explored concepts of a multi-mission modular ship to replace frigates, destroyers, and cruisers. The program suggested that modularity concepts could simplify ship construction while allowing modification of weapon types without major ship alterations. The program developed the variable payload ship (VPS) concept that directly led to the installation of VLS weapon modules on *Arleigh Burke–* and *Ticonderoga*-class ships. Blohm and Voss also used the VPS concept in their popular Mehrzweck-Kombination (MEKO) ship designs.

After a several year gap from the end of the SSES program, NAVSEA initiated the Affordability Through Commonality (ATC) program. ATC recognized the likelihood of future budget constraints and focused on reducing the cost to build and modernize Navy ships through commonality and standardization of equipment. Through several case studies, ATC suggested the implementation of commonality and modularity was an effective method of reducing ship total ownership costs. The program formed the Total Ship Open Systems Architecture (TOSA) team at the Naval Surface Warfare Center Carderock Division (NSWCCD) and influenced many follow-on Navy modularity efforts.

During the same period as the ATC program, the Department of Defense (DoD) established an Open Systems Joint Task Force (OSJTF) to focus on decreasing costs and increasing interoperability and modularity in future combat systems. OSJTF developed a Modular Open Systems Approach (MOSA) guide and rating system as the design standard for future combat systems.

The TOSA team formed during the ATC program developed physical and functional interface standards that constitute the building blocks of ship open system architectures. The TOSA team evolved into the LCS Mission Systems and Ship Integration Team (MSSIT), developing the interfaces between the LCS sea frame and mission module packages.

Recognizing the growing reliance on commercial information technology and how rapidly commercial systems change, NAVSEA started the Open Architecture Computing Environment (OACE) program in 2003. OACE sought to increase combat system software flexibility, motivated by a need for open architecture standards. Program officials selected standards for physical media, networks and protocols, operating systems, middleware, and programming languages. The key principles developed by the OACE team were incorporated into the DoD information technology standards registry and influenced open architecture in the Aegis combat system (ACS).

In 2003, two parallel efforts were initiated, one by NAVSEA, the other by NSWCCD. PEO Ships established the LCS MSSIT to manage module and ship systems development and integration for the LCS program. The NSWCCD Architectures, Interfaces, and Modular Systems (AIMS) program was an evolution of the ATC and TOSA efforts, with a charter to promote increased ship system modularity and interface standards. The effort aimed to facilitate technology refresh and insertion, decrease life cycle costs, and increase mission readiness and flexibility. The outgrowth of this effort was the *Modular Adaptable Ship (MAS) Total Ship Design Guide for Surface Combatants* published by the NAVSEA Engineering Directorate.[3]

These programs and myriad other studies and reports on the topic of modularity have worked from the premise that modularity reduces total ownership costs, hastens new technology insertion, and increases the mission flexibility of naval ships. Separating the construction of the ship from the design and construction of the modules that go on it enables module design and build to proceed in parallel with ship construction. This approach could lead to shorter total ship construction times, also enabling modules to be built in a less-costly location other than the ship construction site. Acquisition costs might also decline if modularity leads to more competition. Modularity tends to reduce both the cost of modernizing in-service ships with new technologies and the time required to upgrade the ship. Shorter modernization periods from the quicker insertion of new technologies would provide more operational ships.

The various studies and reports suggest that modularity is accomplished in various ways. In the next section, we characterize these various forms of modularity and how modularity relates to flexibility.

---

[3] See Karvar et al., 2011.

Different Types of Modularity and Relationship to Flexibility

The Navy efforts described above and other studies of modularity in ship design have led us to define three different types of modularity (see Figure 2.2).[4] We discuss these three types of modularity in the following sections.

Common Modules Across Multiple Ship Classes

All modern civilian and commercial ships are built with construction modules and shipbuilders assemble their ships like a set of toy bricks. These modules are designed for the specific ship class, given the shipyard's construction capabilities, and typically are not considered during the design of other ship classes.

The information technology community has begun efforts to use similar "modules" across multiple classes of ships. Common architectures and common networks are the goal for the majority, if not all,

Figure 2.2
There Are Different Types of Modularity

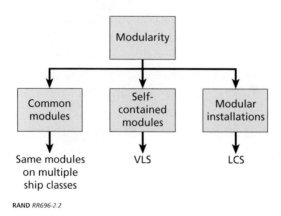

RAND RR696-2.2

---

[4]   Dr. Norbert Doerry has published numerous papers that present examples of modularity applications, some of which we cite in our bibliography. A complete and updated bibliography of Dr. Doerry's papers can be found online at http://doerry.org/norbert/papers/papers. htm (as of February 28, 2013).

naval ships.[5] Many of these modular approaches center on computer hardware and software. Standard racks and interfaces allow for commonality across all ships in a class as well as ships in other classes. The design of the hull, mechanical, and electrical (HM&E) aspects of a ship has been slow in considering modules that could be used across multiple classes of ships.

Many functions on a ship are "hotel-" or personnel-related and have common requirements across classes. Examples include berthing spaces, medical facilities, galleys, and laundry rooms. The size of such facilities may vary across ship classes because of different crew sizes, but the basic tasks and the means to accomplish them are common across all ships. The MAS Guide describes a *pre-engineered element* as "a pre-designed part of the ship that provides space, structural support and services for one sub-function located therein."[6]

Future ship designs could adapt such a modular approach for different hotel functions. As an example, a common space for medical support with all the structural support and power, cooling, and other service interfaces could be designed and required in all future ship designs. The common module that would fit inside that space could be built in a factory-like environment separate from the ship construction site. If a larger crew necessitated bigger facilities, multiple standard modules could be employed. The factory-built module(s), possibly fully outfitted with equipment and tested, could be inserted into the designated space(s) in the ship. This is the approach used to build passenger cabins and other functional areas during cruise ship construction.

Designing and building common modules applicable across multiple ship classes is aimed at cost reductions during ship design and construction. Because of the potential to reduce cost, this type of modularity deserves a broader consideration by the Navy.

---

[5] The Consolidated Afloat Networks and Enterprise Services (CANES) program provided a streamlined network, replacing several previous networks, that will allow a greater level of interoperability in the surface fleet. CANES is currently being installed on multiple classes of ships.

[6] Karvar et al., 2011

Self-Contained Modules

A second form of modularity involves self-contained modules that provide a plug-and-play capability. These modules provide standard connections and interfaces within defined boundaries and are typically used for a specific type of system, such as a missile launch capability or a communication network. Self-contained modules are typically unique to a single class of ships. A primary example is the VLS systems on the *Arleigh Burke* class of destroyers (although it is also used on cruisers). These systems can be loaded with any of several different types of missile systems without changing the form or function of the launcher system. Container ships are an example of self-contained modules in commercial shipping. The ship can carry any number of standard containers (i.e., modules) that can be loaded with practically any type of cargo.

Unlike the concept of using the common modules across multiple ship classes, this form of modularity has been used in Navy programs. In addition to the VLS example, the launch and torpedo tubes on submarines can fire different types of missiles or unmanned systems. Aircraft carriers use this type of modular approach on flight decks and aircraft service areas to accommodate multiple generations of aircraft. Information technology systems adopt this type of modularity by allowing standard racks and interfaces that can accommodate different hardware as technologies evolve.

The advantage of self-contained modules is that they provide a relatively quick and less-costly way of adapting to changes in various payloads, thus reducing the cost of modernizing a ship during its operational life. As long as the payload—for example, a missile or an aircraft—has the necessary interfaces and can fit within the module boundaries, few if any changes need to be made to the ship. Self-contained modules provide flexibility within the constraints of the modules and interface boundaries.

The Navy should continue to study the design and build self-contained, payload-oriented modules for new ship classes. It should also study other potential areas for self-contained modules, such as power generation, electrical distribution, and hotel functions. Self-contained modules have their greatest benefits when components of the module are expected to change over time. If a technology is not expected to change for a significant time, then a self-contained module will have provided flexibility that may not be required. As an example, the torpedo tubes of *Seawolf*-class submarines were designed and built larger than standard torpedo tubes on other submarines in anticipation of larger-diameter weapons that would be needed during the Cold War. The Cold War ended before new, larger weapons were developed. The extra flexibility provided by the larger tubes has yet to be fully utilized.

Modular Installations

A third form of modularity is a modular installation in which the ship (often referred to as a sea frame) can accommodate various types of payload modules through defined interfaces and connections. It is the separation of the ship platform from the systems needed to perform different missions. The interface must be rigidly defined at the beginning and tightly controlled during the life of the ship. A prime example of this type of modularity is the LCS program. There are different payload packages for missions such as anti-submarine warfare (ASW), mine countermeasures (MCM), and surface warfare (SUW). A mission package includes mission modules with mission systems and support equipment, support aircraft (such as the MH-60S), and the crew detachments for the mission modules and aircraft. Mission modules are assemblies consisting of structure, mounted equipment or components, and associated systems that perform an independent function or logical task related to a warfighting mission and are separately testable. Modules have strong interdependencies among their components and limited external interfaces. Some of the attendant mission systems in different modules are housed in standard ISO commercial shipping containers that embark and are secured in the ship. Current MCM mission modules also include vehicles, their ISO-compliant cradles, and support equipment. Other modules are designed to occupy a topside weapon zone (e.g., 30mm gun mission module). The sea frame has space, structure, and connections for these various modules. The ship can perform only one payload package mission at a time but can change that mission by swapping out the modules that comprise a package.

This type of modularity may move the platform from a multi-mission capability, able to perform different missions simultaneously, to a single-purpose "focused mission" platform, such as in the case of the LCS. However, larger platforms like the Royal Danish Navy's *Absalon* class could have enough space to incorporate multiple mission packages that could be switched as needed.[7] The Navy of the future will have both types of ships. The need for future ships that can rapidly switch mission packages was mentioned by the CNO in a recent article.[8] As with the two previous types of modularity, modular installations involve defined boundaries. As long as the module "fits" within those boundaries, the system can function.

Rather than having a costly modernization period for the ship, new modules, with new capabilities, can be developed and built independent of the ship as long as the new modules are compatible with the space and interfaces available on the host platform. Modular installations also allow the development and construction of the payload modules to be decoupled from the design and construction of the platforms that support those modules, thus reducing the time to get new capabilities into the fleet.

There is a question of whether a sea frame with modular installations of mission payload packages will actually realize the full capability of switching payload modules to fit mission needs. The installation of specific mission payload modules on an LCS may become permanent, with those mission modules never replaced. The Royal Australian Navy ANZAC frigate, based on the MEKO design, incorporates several containers for weapons and electronics systems. The ships have undergone several limited-scope modernizations since entering service in 1996, but the containers were neither removed nor replaced during these modernizations. Currently, ANZAC frigates are undergoing a major project to install phased-array radar for missile defense. Con-

---

[7] The Royal Danish Navy's *Absalon*-class ships have a large payload bay with standard connections for power, air, and other services on the walls and floors. Equipment for various missions, such as command and control or humanitarian assistance, is placed in standard shipping containers. The payload bay can accommodate multiple containers that allow the conduct of different missions.

[8] See Greenert, 2012.

currently, the ships' communication systems are receiving an upgrade involving the removal and replacement of certain containers, the first time a container has been replaced for an upgrade.

There are relationships between the three categories of modularity. There are also no clearly defined boundaries; some examples of modularity fit into one or more categories. For example, self-contained modules can both be used for modular installations and be common across multiple ship classes.

Modularity Is Different from Flexibility

All of these forms of modularity allow a ship to adapt to new technologies and missions. But modularity does not necessarily guarantee adaptability, and flexibility is also needed. One way to distinguish between modularity and flexibility in ship design is that modularity sets defined interfaces within rigidly defined boundaries. Flexibility allows both the boundaries and the interfaces to change. A modular platform has a degree of flexibility but also has limits implied by the defined interfaces and fixed boundaries. A truly adaptable ship should not only be modular but also have the ability to expand boundaries and adjust interfaces. This definition of flexibility implies greater space as well as the ability (provided by additional space) to expand power, cooling, bandwidth, and other ship services if and when needed. As shown in Figure 2.3, modularity and flexibility both contribute to the adaptability of a ship.

Flexibility Within a Ship Space

One form of flexibility is the design of spaces whose size and purpose can be changed to meet evolving needs.[9] The flexible infrastructure

---

[9]   Andrew J. Levine, William H. Mish Jr., and Timothy M. Lynch, "Application of Physical Open Systems to Meet Technological Requirements and Capabilities—A Modular Reconfigurable Space," *ASNE Day 2008 Proceedings*, Alexandria, Va.: American Society of Naval Engineers, 2008; Richard DeVries, Andrew Levine, and William H. Mish Jr., "Enabling Affordable Ships through Physical Modular Open Systems," paper presented at *Engineering the Total Ship (ETS) Symposium* 2008, American Society of Naval Engineers, 2008; Deaton,

Figure 2.3
Modularity Is One Way to Achieve Adaptability

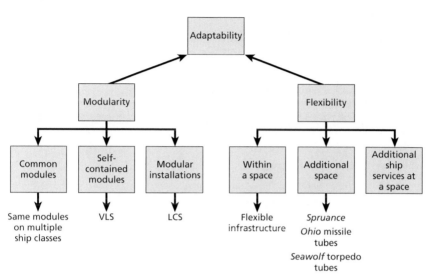

RAND RR696-2.3

being installed on the *Ford* class of aircraft carriers is one example of this type of flexibility.[10] The external bulkheads of a space are permanently located, while the internal bulkheads of the space are mounted on tracks that allow different sizes and configurations of rooms. Standard connections for power, cooling, and computer hook-ups are provided at various locations to accommodate different arrangements of equipment and systems. Also, *Virginia*-class submarines have a reconfigurable torpedo room that can accommodate a large number of special operations forces and all their equipment for prolonged deployments, as well as future off-board payloads.[11]

William A., and James L. Conklin, *Developing Reconfigurable Command Spaces for the Ford-Class Aircraft Carriers*, June 2010.

[10] Appendix C provides a description of the flexible infrastructure on the *USS Gerald R. Ford*.

[11] U.S. Navy, "U.S. Navy Fact Sheet: Attack Submarines—SSN," *Navy.mil*, December 6, 2013.

As mentioned, self-contained modules and modular installations set defined boundaries and interfaces. Flexible infrastructures allow the boundaries of a ship's functional space to change. Rather than requiring substantial hot work[12] to make spaces larger or smaller during a ship's life to adapt to different systems or uses, flexible infrastructures allow quick changes of a space. This type of flexibility provides greater adaptability while reducing the cost of modernizing in-service ships.[13]

Flexibility by Providing Additional Space

Flexibility can also involve designing ships with additional space for expansion of systems or missions. As mentioned above, *Spruance*-class destroyers were designed and constructed with more space and bigger margins than needed and were thus able to serve as the basic ship design for *Ticonderoga*-class cruisers. Also, *Ohio*-class ballistic missile submarines were built with missile tube diameters larger than needed for the C4 missiles in use at the time. These larger tubes provided greater design space for the follow-up D5 missiles, which, although bigger than the C4, fit seamlessly into the missile tubes. A final example is the torpedo tubes of *Seawolf*-class submarines. These tubes had a larger diameter than the standard 21-inch torpedo tubes based on the assumption that the Cold War arms race with the Soviet Union would lead to the need for larger-diameter weapons. Whereas the flexibility in the *Ohio*-class missile tubes resulted in lower upgrade costs, the flexibility provided by the larger *Seawolf* torpedo tube has yet to be utilized.

Aircraft carriers are often cited as truly flexible ships, with the ability to support multiple types of aircraft over their operational lives. CVN-65, the USS *Enterprise*, was recently decommissioned after 50 years of service. The aircraft that flew off the deck of the *Enterprise* ranged from piston engine, propeller-driven A-1s shortly after the ship's commissioning to F-18E/Fs during the recent conflicts in Iraq

---

[12] "Hot work" is any process requiring a fire-watch, such as torch cutting, welding, and grinding. Hot work requires not only the labor to perform the work but also a fire-watch in each affected space.

[13] In many ways, flexible infrastructure for physical spaces is similar to open architectures for information systems.

and Afghanistan. The large size of the Navy's modern aircraft carriers allows this high degree of adaptability. Previously, other large ships had comparable adaptability; for example, the battleship USS *Missouri* used increasingly advanced weapons through the course of its long career, from World War II through the 1991 Gulf War. The Royal Danish Navy's ships of the *Absalon* class have a large, open mission bay with interfaces for ship services placed throughout. Containers designed to support a specific mission are loaded and connected. This is similar in many respects to the LCS concept of mission modularity.

Flexibility Can Involve Providing Additional Ship Services Within a Space

Often, new systems or equipment require more power and cooling than the systems and equipment they replace. A ship may have sufficient services to support the upgraded systems and equipment, but those additional ship services may not be available in the spaces where they are needed. Additional cable, piping, and ducting to the impacted spaces are often needed. Providing more power, cooling, or bandwidth than originally needed to the spaces that hold equipment and systems that have a high probability of being upgraded during a ship's operational life can facilitate the installation of modernization upgrades. Providing additional ship services within a space would involve installing additional electrical cable runs, larger piping, and more fiber cable during ship design and construction.

Accessibility to the equipment and standard connections are two other aspects of a ship design that can improve flexibility. These concepts will be discussed in more detail in subsequent chapters.

Potential Downside to Modularity

One proposed benefit of modularity is the ability to increase a ship class or platform's longevity through periodic modernization efforts.

The U.S. Navy's long-range construction plans[14] assume an expected service life[15] of 40 years for the *Arleigh Burke* class, starting with DDG-79. Figure 2.4 provides a service life assessment for *Arleigh Burke*–class hulls in commission as of May 2012. One can see from the figure that approximately 30 percent of the DDG-51 fleet is more

Figure 2.4
***Arleigh Burke*–Class Service Life Histogram**

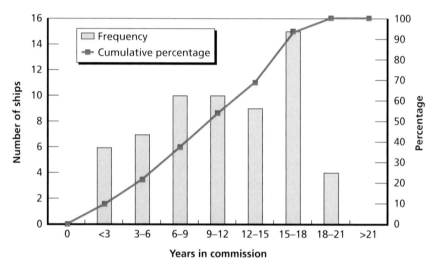

Years in commission

SOURCE: Author analysis of *Naval Vessel Register* data, as of May 2012.
RAND RR696-2.4

---

[14] Office of the Chief of Naval Operations, *Report to Congress on Annual Long-Range Plan for Construction of Naval Vessels for FY 2011*, Washington, D.C.: Department of the Navy, OPNAV N8F, February 2010.

[15] The *Military Equipment Useful Life Study – Phase II, Final Report*, defines *service life* as the total amount of use that the equipment can expend before its capability is degraded to the point where it must be replaced or undergo a major depot-level, procurement-funded service life extension. Service life is a product of the engineering-based original design service life plus any Service Life Extension Program/Recapitalization/Rebuild actions that result in additional capability or additional miles or hours. See Office of the Under Secretary of Defense (Acquisitions, Technology, and Logistics), Property and Equipment Policy Office and Office of the Under Secretary of Defense (Comptroller), Accounting and Finance Policy Office, *Military Equipment Useful Life Study – Phase II, Final Report*, Washington, D.C., May 30, 2008

than 15 years old. We could characterize this fleet as "young" relative to the objective of 30 to 35 years.

Given this young fleet of ships, the *Arleigh Burke* modernization program seeks to extend the service life of the Flight IIA hulls to 40 years.[16] While proponents believe adaptability will help achieve this goal and the Navy's current plans desire the same goal, Figure 2.5 provides insight into an important consideration—extending DDG (and CG) service life to 35–40 years or longer would be a historical achievement for medium-sized surface combatants. There simply is no modern

Figure 2.5
Service Life Assessment of Surface Combatant Programs Since 1960

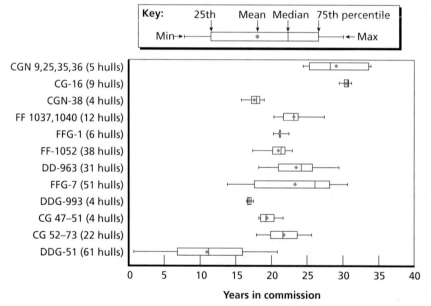

SOURCE: Author analysis of *Naval Vessel Register* data, as of May 2012.
RAND RR696-2.5

---

[16] Ronald O'Rourke, *Navy DDG-1000 Destroyer Program: Background, Oversight Issues, and Options for Congress*, Washington, D.C.: Congressional Research Service, RL32109, 2008.

precedent within the U.S. Navy.[17] Even with a "complete" adaptable design, the bar for extending the usefulness of a platform beyond 30 years is quite high.

In the future, constrained defense budgets may lead to long gaps between the design of new ships and submarines. These long gaps can create problems in sustaining the design resources needed for a new program. If modularity and flexibility concepts allow ships to be modernized at a lower cost more easily, the Navy may keep ships in the operational force longer, thereby exacerbating the problem of long gaps between new design efforts. Also, these adaptability concepts may result in new design efforts involving mostly a repackaging of standard modules and interfaces. This would contribute to the problem of sustaining design resources, especially those that foster new thinking in ship designs.

Current Status of Modularity and Flexibility Initiatives

All of the studies and past efforts on modularity and flexibility are establishing a foothold in naval ship design and construction. The various programs on open architecture for information technology (IT) systems have led to IT environments where new hardware or software technologies and applications are inserted more quickly and at lower costs. The Acoustic Rapid COTS (commercial off the shelf) Insertion (ARCI) program has provided a systematic procedure for IT upgrades on naval submarines. The CANES program is providing a secure and more-easily upgraded IT environment on surface ships.

Modular mission packages using defined interfaces are the backbone of the LCS program and flexible infrastructures are used on the new *Ford*-class aircraft carriers. But the majority of the Navy's modular applications to date have been program-specific, lacking a central focus and direction. The *Modular Adaptable Ship (MAS) Total Ship Design Guide for Surface Combatants* should provide some of that direction.

---

[17] Both aircraft carriers and battleships have lasted for much longer periods, as long as half a century in some cases. Their immense sizes have helped them to accommodate novel payloads over their long lives: the aircraft aboard a carrier have been repeatedly upgraded, while World War II battleships were able to launch 1990s-era missiles.

A stronger proponent of modularity and flexibility within NAVSEA would help to solidify and centralize future modularity and flexibility initiatives.

Three hurdles or questions remain for further modularity and flexibility applications. One hurdle that the Navy must address in deciding how to advance modularity and flexibility concepts is gaining a better understanding of what those concepts may mean in the future. That is, what types of modularity and flexibility will be needed, and what are the best ways to achieve them? That topic is discussed in the Chapter Four. A second hurdle is a better understanding of the cost implications of modularity and flexibility. As described previously, all the studies and reports espouse cost savings, but there have been no definitive studies on total ownership cost effects. Designing to allow for modularity and flexibility typically leads to larger, heavier ships. The current cost-estimation models used during the ship acquisition process are based primarily on weight.[18] These models suggest larger, heavier ships result in greater construction costs. Many argue that ship construction cost-estimating models should be process-based rather than weight-based and that modularity could lead to less-expensive construction processes.[19] The potential cost effects of modularity and flexibility are described in the next chapter. The final hurdle is that there is neither a central proponent for modularity and flexibility nor is there a policy emphasizing modularity and flexibility in ship designs.

---

[18] A Naval Postgraduate School study examining construction costs for naval submarines found that density was also a contributor to submarine construction costs, with less-dense ships resulting in lower costs. See Benjamin P. Grant, *Density as a Cost Driver in Naval Submarine Design and Procurement*, Monterey, Calif.: Naval Postgraduate School, June 2008.

[19] The development of a Product-Oriented Design and Construction (PODAC) cost model was initiated several years ago but was never fully realized. See John C. Trumble, John J. Dougherty, Laurent Deschamps, Richard Ewing, Charles R. Greenwell, and Thomas Lamb, *Product Oriented Design and Construction (PODAC) Cost Model – An Update*, paper presented at the 1999 Ship Production Symposium, July 29–30, 1999; S. N. Garver and J. Edyvane, "Ship Modularity Cost-Reduction Models," *ASNE Proceedings 2010*, Alexandria, Va.: American Society of Naval Engineers, 2010; and Scott C. Truver, "Navy Develops Product Oriented Design and Construction Cost Model," *Program Manager*, Vol. 30, No. 1, January/February 2001.

CHAPTER THREE

How Adaptability Influences Total Ship Life Cycle Costs

The Navy certainly desires adaptable ships with the ability to quickly and inexpensively incorporate new missions and new technologies. Unfortunately, the track record for quickly incorporating needed modernization packages has been spotty at best. Modularity and flexibility are commonly proposed as concepts that contribute to adaptable ships. But what does increased modularity and flexibility cost during the design and construction of a ship? What are the potential savings from modularity and flexibility during a ship's operational life? This chapter addresses the question of the effect of increased modularity and flexibility on a ship's total life cycle cost.

The chapter first discusses how increased levels of modularity and flexibility might affect design and construction costs. Unfortunately, we found no suitable data or analyses that definitively demonstrated any direct relationship between increased adaptability and procurement costs. We do, however, describe the potential effects on design and construction costs of both modularity and increased flexibility. The chapter then addresses how increased modularity and flexibility could affect mid-life modernization costs. Here, we have access to data from the mid-life modernization of the USS *John Paul Jones* (DDG-53). We show the results of some preliminary analyses of those data and suggest what factors drive mid-life modernization costs. The chapter concludes with recommendations on considerations during a ship's design that could reduce the time and cost of mid-life modernizations.

Impact of Adaptability on Ship Design and Construction Cost

Although we found no definitive data or published analyses of the effect of adaptability on a ship's design and construction costs, numerous studies and reports have suggested cost savings as a result of using modularity during ship construction. These studies have been forward-looking, estimating the cost effects of increased modularity based on assumptions of what could happen in the future. Potential cost savings come from reduced construction labor hours, shorter construction periods (by having ship and system design and construction occur in parallel), and the potential for system competition.[1] Most of these reports have considered self-contained modules, such as the VLS on *Arleigh Burke*–class destroyers. As described in Chapter Two, there are various types of modularity, each of which could influence design and construction costs in different ways. Also, flexibility is something over and above those modularity concepts.

How Common Modules Could Affect Design and Construction Costs

A common module is a self-contained portion of the ship structure that is the same across all ships in the class and is built in a factory environment, rather than on the ship while in a dry dock or ship-construction facility. A relevant example is the cabins on cruise ships, which are built outside the shipyard, transported to the ship construction site, and installed in designated spaces within the ship structure. All power, water, and other services from the ship are easily connected to the cabin interfaces. Potential common modules for naval ships could be applications for the hotel functions on the ship, such as galleys, berthing, laundry, and medical facilities. An important advantage of common modules is their potential use not only within a ship class but also across multiple classes of ships. The same hotel module used on a surface combatant could also be used on an amphibious ship.

---

[1] One examination of the cost effects of modular concepts is presented in J. W. Abbott, A. Levine, and J. Vasilakos, "Modular/Open Systems to Support Ship Acquisition Strategies," *ASNE Day 2008 Proceedings*, Alexandria, Va.: American Society of Naval Engineers, 2008.

Greater use of common modules has the potential to reduce design costs—if not necessarily for the first ship using a newly designed module, then certainly for other ship classes that adopt that common module. There are strong arguments for the use of common equipment, such as valves, pumps, or actuators, across different classes of ships. Using common modules is the next step. Design costs are lower when using something already available rather than designing it from scratch. Building common modules in a factory-like environment should lower the construction cost of those modules compared with building them on the ship during construction.[2] Using common modules across ship classes could also reduce construction costs as a result of higher productivity for larger production quantities. Finally, larger production quantities that open the market to firms other than shipbuilders could promote competition, leading to lower costs.

Testing costs should also be lower since the module can be tested and any discrepancies fixed before it is delivered to the ship for installation. The construction of *Virginia*-class submarines use this off-hull testing of a major module with the Command and Control System Module Off-hull Assembly and Test Site (COATS), where the installation, integration, and testing of the combat system is performed before it is loaded into the ship. In fact, each construction module is tested and certified prior to delivery to the assembly shipyard.

In summary, greater use of common modules, especially across ship classes, should lead to lower design and construction costs because of the repeated use of a common design and the building and testing of the common module in a less-expensive, off-ship environment.

How Self-Contained Modules Could Affect Design and Construction Costs

Self-contained modules typically apply to a single class of ships and involve defined connections and interfaces within defined boundaries.

---

[2] There is a widely mentioned "8-3-1" rule-of-thumb in shipbuilding: a task that takes one hour in a shop requires three hours on the platen and eight hours on the ship in the dry dock. See John F. Schank, Hans Pung, Gordon T. Lee, Mark V. Arena, and John Birkler, *Outsourcing and Outfitting Practices: Implications for the Ministry of Defence Shipbuilding Programmes,* Santa Monica, Calif.: RAND Corporation, MG-198-MOD, 2005.

They are used for a specific application, such as a weapon system or a communications network. Self-contained modules are typically used when there is a high probability that technologies within the system will change. They provide a relatively quick and less-costly way of adapting to technology changes during the life of the ship. Therefore, self-contained modules aim to reduce ship modernization costs rather than reduce initial design and construction costs.

There is certainly a debate over the effect that self-contained modules have on design and construction costs. If the standards and interfaces have been previously defined, design costs could be lowered. If, however, new standards and interfaces are needed for the self-contained module, design costs could be higher. It is difficult to project the degree to which self-contained modules affect construction costs. Again, if the module is adopted from other ship classes, construction costs could be lower as a result of economies of scale and increased learning when building the module. If the module is being built for the first time, construction costs could be higher if the standards and interfaces involve more structure or connections to incorporate modularity.

It is also difficult to project the effect on design and construction costs of self-contained modules. The Navy should continue to explore the use of such modules on specific classes of ships and should try to use the same modules across ship classes. To some degree, this is the approach the IT community is using with CANES installation on multiple ship classes.

How Modular Installations Could Affect Design and Construction Costs

Modular installations have much in common with self-contained modules. Where self-contained modules are aimed at a specific system, such as missile launching, modular installations allow various types of systems to be connected to the defined interfaces and connections. The prime example of modular installations in the U.S. Navy is the LCS, where different mission systems can be installed on the sea frame based on the desired mission capabilities of the ship. Mission payload modules can be switched when the ship needs to perform different missions.

The Royal Danish Navy's *Absalon* class of ships is another example of the use of modular installations.

Modular installations separate the design and construction of the sea frame from that of the mission modules. This approach permits parallel development and construction efforts resulting in a quicker introduction of the ship into the fleet. As with self-contained modules, modular installations should have their biggest effect on modernization costs during a ship's operational life. The design and construction costs of the basic ship should be lower since the costs associated with mission systems are not included. However, the mission modules require their own design and construction programs. Therefore, it is unclear whether modular installations increase or reduce the design and acquisition costs of the total operational package.

How Flexible Spaces Could Affect Design and Construction Costs

The best example of this type of flexibility is the flexible infrastructure system for the command and warfare system spaces on the USS *Gerald R. Ford*, the lead ship in a new class of aircraft carriers. During the design and construction of previous aircraft carriers, and almost all other types of naval ships, these spaces would be rigidly defined and walls, overheads, and partitions would be permanently welded to the ship. Electrical and other service connections would be provided at designated locations. This process is very similar to that of house construction, but with the need for welding instead of nails to assemble the structure.

Responding to a requirement for reconfigurable spaces, the Navy and Newport News Shipbuilding, a division of Huntington Ingalls Industry, developed a system of tracks and standard connections that allow the size and configuration of the *Ford's* command and warfare system spaces to adapt and change as needed.[3] The flexible infrastructure minimizes the need for welding (i.e., hot work) during the construction of the ship and practically eliminates welding if the spaces

---

[3] Appendix C provides a description of the components used in the flexible infrastructure. An excellent discussion of the use and development of the flexible infrastructure on *Ford* can be found in Deaton and Conklin, 2010.

are reconfigured during the life of the ship. Hot work requires qualifications, training, and inspection as well as oversight of adjacent areas during the welding process. A business case analysis[4] estimated that material costs for the flexible infrastructure system were greater but that installation labor costs were significantly reduced. The analysis concluded that the design and acquisition costs of the conventional and modular approaches were approximately equal based on the accuracy of the models and assumptions. However, the analysis suggested the costs to reconfigure the spaces were reduced by approximately 25 to 50 percent because of reduced labor associated with hot work. It takes approximately seven years to build an aircraft carrier and numerous changes to the configuration of spaces can occur during that lengthy construction period. These changes typically result in additional cutting and welding to move bulkheads and ship service connections. A flexible infrastructure allows the reconfiguration of spaces during ship construction without the additional cutting and welding, thus reducing construction costs.

The use of flexible spaces during ship design should reduce labor costs, both during the construction of the ship and when spaces are reconfigured during the ship's operational life. A greater use of the same or similar flexible infrastructure components (e.g., tracks, foundations, and connections) should lower material costs during construction. In all, flexible spaces are a cost-effective way to add adaptability to certain areas of a ship.

How Additional Space and Additional Ship Services Could Affect Design and Construction Costs

As discussed in Chapter Two, our concept of flexibility—as compared to modularity—involves bigger spaces and larger margins. Designing a ship with more space than immediately needed and the ability to easily add power, cooling, and other services to a ship space provides a level of adaptability greater than what results from strict modularity. However, Navy ship construction estimates are based primarily on the weight of the ship, so the additional size and weight associated with more

---

[4] Described in Deaton and Conklin, 2010.

modular, flexible ships comes at a cost.[5] The basic trade-off is between potentially greater construction costs for the additional space and margins and the reduced costs of modernizing the ship in the future. This trade-off is not well understood.

Maintaining adequate power and cooling margins to accommodate future growth is also difficult. The margins set for new construction are typically reduced early in a design when cost targets are threatened. Adding power and cooling to an existing ship when adequate margins are not available is very costly, if at all feasible.

Effect of Adaptability on Mid-Life Modernization Costs

A modular and flexible ship should, by definition, require less time and money to modernize equipment and systems during a ship's operational life than a ship with less modularity and/or flexibility. Not only are modernization costs reduced by avoiding costly hot work and structural changes, but the modernized ship can also return to the operational fleet more quickly.

Understanding how modularity and flexibility influence the cost of modernizing naval ships during their operational lives is difficult. The actual costs of changes to a ship's structure and service connections are typically hard to separate from the cost of a total modernization package. Even when these installation costs can be segregated, it is difficult to estimate how those costs could have been reduced if greater degrees of modularity and flexibility were used during a ship's design and construction.

An examination of the *Arleigh Burke* class's modernization experiences and plans provides a better understanding of the potential savings of modularity and flexibility during a ship's operational life and what drives the various elements of modernization costs. The cost to modernize a ship includes the cost of the new equipment, the cost to install and test new equipment, and the cost to design and plan the

---

[5] Grant, 2008, suggests weight-based cost models could benefit from including ship density.

modernization process. Modularity and flexibility will not necessarily result in cost savings for the new equipment but should lead to lower design, planning, installation, and testing costs. The greatest potential cost savings come from the reduced engineering and construction labor to install and integrate the new equipment.

To help understand what factors drive the cost of a mid-life upgrade, we obtained data for the USS *John Paul Jones* (DDG-53) mid-life modernization package. The data were provided by NAVSEA PMS 400F, Bath Iron Works (the DDG-51 Planning Yard), and the Supervisor of Shipbuilding at Bath Iron Works. Various data were provided for the 29 requirements in the modernization work package and included the following:[6]

- requirement number
- description of the requirement, including effect on services
- ship compartments or spaces affected by the requirement
- design man-hours
- production man-hours
- testing man-hours
- government-furnished materiel (GFM) costs
- whether removal was needed
- whether there was an effect on the electrical system
- whether there was an effect on the ship's vertical center of gravity (termed KG).

Table 3.1 shows the 29 requirements.[7] The table also categorizes the type of modernization—whether the requirement dealt with computer hardware or software that was mission-related or ship-related or

---

[6] The requirement number, description of the requirement, and space affected for the 29 requirements are provided in Appendix D.

[7] Details regarding design and production/testing man-hours and the cost of GFM are proprietary. Authorized persons can request a version of this report containing this data by contacting the U.S. Navy's PEO Ships.

Table 3.1
Requirements for USS *John Paul Jones* (DDG-53) Mid-Life Modernization

Requirement Number	Application	Requirement Number	Application
77615	Computer Mission	79256	Computer Mission
78511	Computer Mission	77829	HM&E
73088	Computer Ship	77419	HM&E
77052	Computer Ship	76974	Computer Ship
71604	Computer Ship	76253	HM&E
71605	HM&E	76186	HM&E
78391	Computer Mission	75928/82635	HM&E
74012	HM&E	78513	Computer Mission
71726	Computer Ship	73622	HM&E
76869	Computer Ship	77269	HM&E
78512	Computer Mission	77259	Computer Ship
79584	Computer Mission	76648	HM&E
70403	HM&E	76034	HM&E
77427	Computer Ship	76829	Computer Ship
78819	Computer Mission		

whether the requirement addressed HM&E (i.e., non-computer) aspects of the ship.[8]

Table 3.2 shows the number of requirements and the average and median design man-hours, production/testing man-hours, and GFM cost for the 29 requirements and for each type of modernization. Some observations from the data include the following:

---

[8] HM&E changes may include electronics but do not include the decisionmaking capabilities inherent in a computer system.

Table 3.2
Summary Data for Three Categories of USS *John Paul Jones* (DDG-53) Mid-Life Modernizations

		Type of Modernization		
	Total	Computer Mission	Computer Ship	HM&E
Number of requirements	29	8	9	12
Average design man-hours	1,463	2,566	1,588	633
Median design man-hours	680	1,185	780	585
Average Production/Testing man-hours	17,397	29,219	20,297	7,297
Median Production/Testing man-hours	8,292	17,620	18,728	2,260
Average GFM (thousands)	$268	$250	$366	$207
Median GFM (thousands)	$155	$136	$273	$70

- The average design and production/testing man-hours for computer-related upgrades are approximately two to four times higher than those for HM&E upgrades.
- The average design man-hours and production/test man-hours are greatest for requirements that involve upgrades to the computer hardware and software associated with the ship's missions. However, the average production/testing man-hours for mission computer upgrades are skewed by two high-value requirements (77615 and 78511). The median production/testing man-hours for mission computer and ship computer upgrades are very similar.
- GFM costs ranged from a low of $6,000 to a high of over $1.3 million. The averages across the three types of modernizations were very similar.

These data suggest computer-related upgrades should be the prime target for reducing mid-life modernization costs.

All 29 requirements in the modernization package involved some type of removal and affected the electrical load and the KG. Twelve of the 29 requirements accounted for almost 90 percent of the total work package production man-hours. We requested further information on those 12 requirements, including the number of new foundations, the feet of new cable installed, and the feet of new fiber cable installed. We also asked for a breakout of testing hours from the early production/testing data.[9] The resulting data for the 12 requirements are shown in Table 3.3. Five of the top 12 were computer hardware or software upgrades to the ship's missions, five were for computer upgrades to the basic ship operations, and two were upgrades to the ship's HM&E.

Table 3.3
Data for Top Man-Hour Requirements for USS *John Paul Jones* (DDG-53) Mid-Life Modernization

Requirement Number	Foundations (Number)	Cable (feet)	Fiber (feet)
77615	35	55,231	32,368
78511	73	44,654	8,919
73088	16	18,833	14,304
77052	6	20,926	2,266
71604	8	37,366	0
71605	16	3,192	0
78391	17	10,816	7,362
74012	6	2,626	0
71726	20	16,395	700
76869	4	28,730	0
78512	7	11,497	1,557
79584	1	18,110	5,571

[9] The testing hours were extracted from the ship installation drawings and reflect installation tests (hydrostatic testing, static test pulls, etc.). They do not account for the test hours associated with new equipment. The system/equipment test hours are dependent on the equipment installed and can be significant for more complex systems. We keep the test hours in our analysis but caution on any observations drawn from the analysis of these values.

From a statistical perspective, there are four variables we are interested in estimating: design man-hours, production man-hours, testing man-hours, and cost of GFM. We have variables that may help in explaining these man-hours and costs—number of foundations, feet of cable, feet of fiber, and type of modernization. The matrix of correlation between these variables is shown in Table 3.4.[10] Some observations from the correlation matrix include the following:

- Design hours have a very strong direct linear relationship ($r = 0.84$) with production hours. Design hours also have strong to moderate linear relationships with the number of foundations ($r = 0.84$), feet of cable ($r = 0.73$), and feet of fiber ($r = 0.46$). The type of modernization does not appear to have a linear relationship with design hours.
- Production hours have a strong to medium linear relationship with testing hours ($r = 0.69$), number of foundations ($r = 0.69$), feet of cable ($r = 0.78$), and feet of fiber ($r = 0.81$). The type of modernization does not appear to have a linear relationship with production hours.
- Testing hours have a moderate linear relationship with GFM cost ($r = 0.54$), the number of foundations ($r = 0.45$), feet of cable ($r = 0.65$), and feet of fiber ($r = 0.55$). HM&E modernizations result in fewer testing man-hours compared to computer-related modernizations.
- None of the variables appear to have a linear relationship with GFM cost.
- There are moderate linear relationships between the number of foundations and the feet of cable ($r = 0.57$) and between the feet

---

[10] The correlation coefficient (r) shows the strength of the linear relationship between two variables or the degree of variation of the observations from a straight line fitted to the data. Correlation coefficients range from -1 to $+1$, with -1 indicating a perfect inverse linear relationship and $+1$ indicating a perfect direct linear relationship. Values close to 0 indicate little or no linear relationship. Caution must be taken with the assumptions drawn from correlation coefficients. Correlation indicates how well the data fit a linear relationship but does not suggest the cause of the relationship.

Table 3.4
Correlation Matrix

	Design Man-Hours	Production Man-Hours	Testing Man-Hours	GFM Cost	Foundations	Feet of Cable	Feet of Fiber
Design Man-Hours	1.00						
Production Man-Hours	0.84	1.00					
Testing Man-Hours	0.59	0.69	1.00				
GFM Cost	0.46	0.41	0.54	1.00			
Foundations	0.84	0.69	0.45	0.18	1.00		
Feet of Cable	0.73	0.78	0.65	0.40	0.57	1.00	
Feet of Fiber	0.46	0.81	0.55	0.14	0.44	0.64	1.00
Computer mission	0.27	0.27	0.05	-0.27	0.44	0.31	0.47
Computer ship	-0.11	-0.12	0.28	0.11	-0.29	0.11	-0.25
HM&E	-0.21	-0.20	-0.43	0.21	-0.15	-0.56	-0.30

of cable and fiber (r = 0.64).[11] This suggests that both related variables may not be needed in a linear regression model.

Predicting Design Hours

The best linear regression fit to the design man-hour data is captured by the following formula:

$$\text{Design man-hours} = 484.45 + 71.42\, Fo + 0.05\, C,$$

where Fo is the number of foundations and C is the length of cable in feet.

This regression model has an adjusted coefficient of determination of 0.74, suggesting a good fit of the regression line to the data observations.[12] Both the number of foundations and the feet of cable are significant in the prediction of design man-hours, with each additional foundation (when feet of cable is held constant) adding a little over 71 hours to the design effort and each additional 1,000 feet of cable (when the number of foundations is held constant) adding 50 hours to the design effort.

Predicting Production Hours

The best linear model fit to the production man-hour data is the following:

$$\text{Production man-hours} = 13{,}764.17 + 348.97\, Fo + 0.38\, C \times 1.16\, Fi,$$

---

[11] Correlation between two independent variables can lead to problems with multicollinearity in regression analyses.

[12] The coefficient of determination (R2) is the proportion of the total variation in the dependent variable (here design man-hours) that is explained by the linear regression equation. It reflects how well the regression model fits the observed data for the dependent variable used to generate the model. Adding predicting variables to the regression model will typically lead to an increase in R2 but can result in a model that has predicting variables that are not really helpful in estimating the dependent variable. The adjusted coefficient of determination (R2ADJ) incorporates the effect of including additional predicting variables in a multiple regression equation and is a better measure for a multiple regression model. The adjusted coefficient of determination may increase or decrease as variables are added to the model.

where Fo is the number of foundations, C is the length of cable in feet, and Fi is the length of fiber in feet.

This regression model has an adjusted coefficient of determination of 0.78, suggesting that the model is a good fit to the production man-hour data. Each additional foundation (with the feet of cable and fiber held constant) increases production effort by almost 350 hours. As mentioned previously, the feet of cable and the feet of fiber are strongly correlated with each other. The resulting multicollinearity does not affect the predictability of the overall regression model but does preclude a proper interpretation of the contribution of those variables to production man-hours.

Predicting Testing Hours and GFM Cost

As suggested by the correlation coefficients shown in Table 3.4, we were unable to develop any suitable linear regression models for predicting testing man-hours or GFM cost. There is also the caveat expressed previously that the test hour data do not include the test hours for new equipment, which can be significant for complex equipment. We note that certain systems, such as the ACS, are tested in a factory environment before shipment to the shipyard for installation on the ship. Once the system is installed, additional testing occurs. It would be worthwhile for the Navy to examine this overall testing process to determine if there are any duplicative procedures that could be eliminated to reduce the testing workload for new equipment and systems.

Data Caveats

It is important to note that the data on DDG-53's mid-life modernizations are estimates, not actuals. Although the estimates of the number of foundations and feet of cable and fiber should be fairly close to actual values, the estimates of the various man-hour data may be very different from the actual hours realized during the modernization effort. Also, as mentioned above, the hours provided by the Planning Yard do not reflect all the man-hours associated with the modernization. The hours especially do not include the efforts spent by the Alteration Installation Teams (AITs) when installing and testing new Space and Naval Warfare Systems Command (SPAWAR) equipment.

Finally, there are only data on the single mid-life modernization. Better relationships that estimate design and production man-hours could be developed with actual and complete data from a number of modernizations. Such a data collection effort should be undertaken by the Navy to help better understand what drives modernization costs.

Potential Ways to Reduce Mid-Life Modernization Costs

The available data on a mid-life modernizations and discussions with the DDG-51 Planning Yard suggest there are a number of steps that could be taken to reduce the man-hours required for modernizations. These include the following, each of which is discussed in some detail:

- Improve access to modernized equipment
- Minimize the number of foundations changed during a modernization
- Minimize the amount of new cable and fiber during a modernization
- Increase power, cooling, and data exchange
- Accomplish more pre-installation testing
- Improve planning before modernization
- Coordinate modernizations.

Improve Access to Modernized Equipment

Existing equipment replaced during a modernization must be removed from its location in the ship before the new equipment is installed. For larger, heavier equipment this may require removal of doors, ladders, and bulkheads and may even require cutting holes in the side of the ship. These types of structural changes typically require hot work leading to additional man-hours for welding, removing and replacing the installation, and the oversight in adjacent spaces. In addition to ship structure, interference from ship systems, such as piping, ventilation, or cable runs, may require removal involving additional time and labor. Modernization costs could be reduced if a ship design provided easier access to any major equipment that has a reasonable expectation of being replaced during the life of the ship. The design objective would be to minimize structural and support system changes.

The desire for ship survivability can hinder the access to major equipment. Lessons learned from the loss of the British Royal Navy's HMS *Sheffield* during the Falklands War, the Iraqi jet attack on the USS *Stark*,[13] and the Iranian mine damage to the USS *Roberts* reaffirmed the importance of a number of DDG-51 design changes to increase the ship survivability. Based on the experience of these ships, several critical combat information and combat system spaces were relocated in the steel hull with outboard passageways adjacent to the spaces. The design was to protect the critical spaces from fragmentation weapons and shaped charges. However, these survivability features make modernization of the equipment located within them more costly.[14] The ship design must consider these trade-offs between survivability and equipment access.

A ship design should also consider the placement of ship service systems, such as cooling, power, and ventilation. To the degree possible, these support systems should be situated to minimize any interference during equipment removal and replacement. Also, the design should consider standard connections for the support systems that run between adjacent spaces. Flexibility involving larger spaces and flexible infrastructures are ways to improve accessibility and reduce system interferences leading to potentially lower equipment installation costs.

Minimize the Number of Foundations Changed During a Modernization

Ship equipment is typically mounted on specially designed foundations to support it when the ship is underway and to provide a measure of protection against shock. Currently, the foundation for a piece of equipment being replaced during a modernization will not work with the new equipment. The old foundation must be removed and a new

---

[13] On May 17, 1987, an Iraqi jet fired missiles at the USS *Stark*, resulting in over 50 casualties.

[14] Nine of the top 12 modernization packages for the USS *John Paul Jones* involved these critical combat information and combat equipment spaces.

one installed. This again typically involves hot work with all the associated man-hours.

Our preliminary analysis of the USS *John Paul Jones* mid-life modernization data suggest that the number of new foundations needed for the modernization directly influences design and production man-hours. To the extent possible, standard foundations for specific types of equipment should be used during ship design. This is one objective of the flexible infrastructure on the USS *Gerald R. Ford*. Replacement equipment should be designed to fit on these standard foundations.

For in-service ships, any new equipment installed during a modernization should be designed to fit on existing foundations to the maximum extent possible. When approving future ship alterations, a cost trade-off analysis is warranted between building new equipment to fit on old foundations versus building and installing new foundations.

Minimize the Amount of New Cable and Fiber During a Modernization

Similar to the number of foundations, the more cable and fiber required during a modernization, the higher the installation costs. Most computer-related modernizations require tens of thousands of feet of new cable and fiber. HM&E modernizations rarely require new fiber but can involve thousands of feet of cable. Higher power margins during ship design, additional electrical capacity and connections, and new equipment that utilize existing cable and fiber can all help reduce modernization costs.

Increase Power, Cooling, and Data Exchange

As ship systems become more complex, they typically require additional electrical power. Some ship systems that were previously powered by hydraulics are being converted to electrical power. New weapons are envisioned that use large amounts of electrical power, and the new aircraft launch system on *Ford*-class aircraft carriers uses electric power instead of steam. Future ships will likely require substantially more power than the current ships in the fleet. As the need for electrical power increases, so does the need for additional cooling. Also, stand-alone systems and equipment are being replaced with more integrated systems, and larger amounts of data are being collected and transmitted throughout the ship.

New ship designs recognize this need for additional power, cooling, and data transfer during a ship's operational life. The *Ford* class has a new power-generation system that provides almost three times the capability of the older *Nimitz* class. More power and cooling are being considered for the DDG-51 Flight III ships than is immediately needed. Future ship designs must recognize this trend and incorporate substantial margins for ship support systems. The key is having enough power, cooling, etc., to sustain the ship during its service life, not just to add ship services for the sake of having "more."

Accomplish More Pre-Installation Testing

New equipment installed during a modernization requires testing to ensure the equipment works as planned. This testing is especially critical for integrated systems that involve various computer hardware and software components. Several ship programs have used pre-installation of equipment and systems to help reduce costs due to rework or modifications after installation. The *Virginia* class uses the COATS facility to test the combat system thoroughly before it is inserted into the submarine. The Aegis cruiser program used a Unitized Foundation Project to provide foundations to the equipment supplier to permit installation, integration, and testing before the combat system was delivered to the shipyard for installation on the ship.

New ship programs should involve pre-installation assembly and testing of equipment and integrated systems prior to delivery of the equipment or system to the shipyard. This concept should be used not only for new ship construction but also for any major modernizations during the life of the ship.

Improve Planning Before Modernization

Discussions with the activities that plan and execute major modernizations suggested better planning and execution of the modernization could also lead to reduced costs. Currently, there are separate planning processes for repair work and modernization work needed during a ship's visit to a shipyard for an availability. For *Arleigh Burke*–class ships, the Planning Yard develops the modernization work packages and the NAVSEA Surface Engineering Maintenance Planning Program develops the repair work packages. Typically, these two work package preparation processes do not interact. The two-step process is likely duplicative and adds cost. Coordination could help to integrate the scheduling and conduct of the two types of work to minimize interferences and conflicts and to take advantage of common activities. This requires a single Navy-directed manager to merge the repair and modernization work packages. The Navy is currently planning on merging the two processes in fiscal year 2015 with full integration of the two planning processes by fiscal year 2016.

Our discussions with the shipyards that execute the availabilities suggest there are various configurations for the different ships within a class or flight and that the documentation provided by the Navy is typically not consistent with the actual ship configuration. This mismatch in ship configurations leads to additional rework because equipment and modules designed for one configuration do not exactly match the actual configuration of the ship.

The Navy submarine force has established disciplined management of all facets of submarine maintenance and modernization. It maintains complete configuration of the submarines and effectively manages modernizations and repair. This strict control of ship configurations should be used by all ship classes, especially for large classes of complex ships. Also, photogrammetric studies prior to availabilities, especially in critical spaces, can be an effective tool for defining a ship's configuration.

Coordinate Modernizations

Similar to the planning process for a ship's availability, the execution of the repair and modernization packages are not coordinated. For example, many modernization work packages that are computer related are accomplished by AITs. These teams focus on specific capability upgrades. However, AITs are not always integrated into the executing shipyard's schedule and can disrupt the scheduled repair work. It is not unusual for an AIT to work on a space after the repair work in that space has been accomplished and the space closed out to further work. The AIT then comes in, and some of the preparation, installation, and testing must be done again. Coordinating both the planning and execution of repair and modernization work can help eliminate or reduce extra work.

CHAPTER FOUR

Technological Trends, the Geopolitical Context, and Historical Lessons

Historical experiences can help to shape future decisions regarding ships' flexibility with respect to changing mission sets or new technologies in two ways. The first is by providing trend data that can be used for projections. While trends can change or even reverse over time, thoughtful assessment of past trends can provide insight into future developments. Specifically, analysis of technological and geopolitical trends can aid in anticipating both future mission sets and the ship requirements associated with those missions. The second way history can inform decisions is by providing examples of earlier attempts to make ships more accommodating of new mission sets or novel technologies.

This chapter examines these two applications of historical experience. It begins by reviewing technological and geopolitical trends, examining what these likely portend in terms of future flexibility requirements. The second portion of the chapter turns to how ships accommodated novel missions and technologies during a time of particularly rapid change in both, from the mid-19th to the mid-20th centuries.

Technological Trends

The following four key technological trends appear likely to have a considerable influence on the ways in which ships will operate over the next several decades:

- The rapidly increasing use of off-board unmanned systems
- The growing importance of using of the electromagnetic spectrum as a weapon
- Enhanced capabilities for long-range targeting
- The increasingly networked nature of the battlespace.

We discuss these four trends below.

Unmanned Systems

Unmanned military platforms will likely have a considerable influence on warfare over the next several decades. Advances in electronics, IT, communications, robotics, materials science, mechanical engineering, energy storage, and other fields are contributing to the ability of unmanned systems to perform missions more effectively, at lower risk, and at lower cost than manned platforms.

The development and use of unmanned platforms have increased enormously over the past decade, particularly unmanned aircraft systems (UASs), which are increasingly used by all four military services. (Some services refer to these a remotely piloted vehicles, or RPVs.) As recently as 2002, the entire DoD had only 167 UASs; by 2010, this number had increased to 7,500.[1] Initially, the vast majority of UAS missions were associated with intelligence, surveillance, and reconnaissance (ISR). (Incidentally, the same was true of manned military aircraft during their early development.) UASs have since diversified to support a wider array of roles, including ground targeting and communications support. The types of UAS programs have also diversified considerably in recent years, to include aircraft of dramatically different sizes and capabilities. The *Unmanned Systems Integrated Roadmap FY2011–2036*, published in 2011, lists 21 UAS programs.[2]

While less visible than UAS programs, those associated with unmanned surface vehicles (USVs) and unmanned undersea vessels

---

[1] Jeremiah Gertler, *U.S. Unmanned Aerial Systems*, Washington, D.C.: Congressional Research Service, R42136, January 3, 2012.

[2] Department of Defense, *Unmanned Systems Integrated Roadmap FY2011–2036*, Washington, D.C., 11-S-3613, 2011a, p. 18.

(UUVs) have also been growing rapidly.[3] The previously mentioned *Unmanned Systems Integrated Roadmap FY2011–2036* includes six USV programs and 13 UUV programs.[4] Applications include MCM, ISR, ASW, oceanographic survey, explosive ordnance disposal, force protection, and target acquisition. RAND has conducted three separate studies on how the U.S. Navy can effectively employ USVs, UUVs, and UASs in a variety of contexts.[5]

The growing use of unmanned systems can affect warship requirements in terms of personnel, space, power, and communications. We discuss each of these below.

Personnel

It might be expected that not having people aboard these systems would reduce the number of personnel associated with them aboard the ship on which they are based. However, one of the paradoxes of "unmanned" systems is that they typically require a great many people to maintain and control them, as well as interpret the sensors of such systems. Operating an unmanned system often requires one person to control the vehicle, and another to control the vehicle's payload or monitor sensor outputs, such as streaming video.[6] For many missions, an imagery analyst is also required. The protracted nature of many unmanned missions implies that multiple shifts of personnel may be

---

[3] Unmanned ground vehicles (UGVs) have also experienced dramatic increases in usage and diversity. However, these are unlikely to be launched directly from U.S. Navy ships.

[4] Department of Defense, 2011a, p. 20.

[5] See Brien Alkire, James G. Kallimani, Peter A. Wilson, and Louis R. Moore, *Applications for Navy Unmanned Aircraft Systems*, Santa Monica, Calif.: RAND Corporation, MG-957-NAVY, 2010; Scott Savitz, Irv Blickstein, Peter Buryk, Robert W. Button, Paul DeLuca, James Dryden, Jason Mastbaum, Jan Osburg, Philip Padilla, Amy Potter, Carter C. Price, Lloyd Thrall, Susan K. Woodward, Roland J. Yardley, and John M. Yurchak, *U.S. Navy Employment Options for Unmanned Surface Vehicles (USVs)*, Santa Monica, Calif.: RAND Corporation, RR-384-NAVY, 2013; and Robert W. Button, John Kamp, Thomas B. Curtin, and James Dryden, *A Survey of Missions for Unmanned Undersea Vehicles*, Santa Monica, Calif.: RAND Corporation, MG-808-NAVY, 2009.

[6] Harlan Geer and Christopher Bolkcom, *Unmanned Aerial Vehicles: Background and Issues for Congress*, Washington, D.C.: Congressional Research Service, RL31872, November 21, 2005; UAV GROUND CONTROL STATION (GCS) Basis of Issue Plan, undated.

necessary. As one indication of the person-to-machine ratio, the Indian Navy is inaugurating a UAS squadron that will have 12 officers and 50 sailors maintaining and operating four unmanned aerial systems.[7] While the person-to-machine ratio may be reduced by increasing machine autonomy as well as automated analysis of sensor outputs, the effect of this is unclear given the desirability of using human judgment to make key decisions. As new systems are designed, the hope is that integrating manned and unmanned operations and maintenance will reduce shipboard manpower.[8] One vision for the future UAV squadron is a traditional pilot with knowledge of how weather and mechanical issues affect real-time flight who supervises and makes key decisions for multiple UAVs while supported by teams of sailors who would monitor actual flight operations.[9]

An operational example of a USV is cited in a 2006 Naval Post-graduate School thesis. The ISR-configured Spartan Scout assigned to the USS *Gettysburg* (CG-64) had a USV team consisting of 18 personnel. Boatswain's mates and seamen launch and recover the Spartan Scout, as they also do for the manned Rigid Hull Inflatable Boat (RHIB). The USV requires a minimum of four personnel to operate it: one to operate the remote operating station, a command and control operator to monitor sensor displays, a radio control operator to control during launch and recovery, and a coxswain for manned operations, plus personnel for electronic and mechanical repairs. Some of the *Gettysburg's* personnel were informally trained to operate and maintain the Spartan Scout before deployment, while other personnel received training on board. Handling and maintenance responsibilities for the

---

[7]   See C. Jaishankar, "UAV Squadron to Come Up to Uchipuli," *The Hindu*, April 8, 2012. The article does not mention how or if the officers and sailors will have other shipboard duties.

[8]   Lynden D. Whitmer, *Naval UAV Programs: Sea Based UAV's*, Dahlgren, Va.: Naval Surface Warfare Center Dahlgren Division, February 26, 2002, p. 3.

[9]   Andrew Tilghman, "New Rating Considered for UAV Operators," *Navy Times*, November 2, 2008.

USV were assigned to officers and diverse enlisted ratings, depending on the skill sets required.[10]

Personnel requirements are exacerbated by many unmanned systems' ability to operate for long periods, outlasting individuals' capabilities to control them effectively and monitor their outputs without fatigue degrading performance. For this reason, multiple shifts of personnel may be associated with a single vehicle. Routine maintenance and repair add to overall personnel requirements.

The growing capabilities of unmanned aircraft are likely to increase demand for personnel in the near term: as unmanned systems' endurance, sensor capabilities, and design complexity increase, more personnel will be required to support each mission. In addition, the relatively small sizes of many unmanned platforms relative to their manned counterparts may mean that more are being deployed from each ship, correspondingly increasing the number of personnel required. Some of this burden may be alleviated by technological advances that reduce maintenance requirements, increase autonomy, and enable automated interpretation of sensor outputs. Also, not all of the individuals involved in supporting an unmanned system need to be aboard the ship from which it was launched. Some control functions and sensor interpretation could be done by personnel ashore or in the air, although that would require extended communications links. The complexity inherent in such an arrangement would create opportunities for breakdowns or miscommunications, or time lags, particularly in the face of adversary interference or battle damage.

On balance, then, there are good reasons to suspect that the increasing use of unmanned systems will not result in markedly fewer personnel being aboard warships anytime soon. While the precise personnel requirements relative to today's manned platforms are unclear, warships intended to last for decades should be designed with the anticipation that personnel requirements to support missions will not necessarily decline, and may even grow. Personnel, in turn, require

---

[10] Wayne Galye, *Analysis of Operational Manning Requirements and Deployment Procedures for Unmanned Surface Vehicles Aboard US Navy Ships*, thesis, Monterey, Calif.: Naval Postgraduate School, March 2006, p. 11.

space, power, and other resources for support. Ideally, all shipboard aircraft would be as compatible as possible with existing maintenance facilities to minimize integration problems and adhere, where possible, to requirements specified by the Naval Aircraft Maintenance Program (NAMP).[11]

Power

All unmanned systems consume power. Whereas manned platforms typically consume fossil fuels, necessitating fuel storage on the ship, unmanned systems include a mix of battery-powered and fuel-consuming vehicles. The *Unmanned Systems Roadmap 2007–2032* lists 13 current and proposed Navy and Marine Corps unmanned vehicles; ten are powered by liquid fuels and three by batteries.[12] The smaller size of many unmanned systems compared with their manned counterparts and the lack of need for support systems for humans will likely result in less consumption of either fuel or electrical energy relative to manned systems (though this may be offset by the greater number of unmanned systems). Unmanned systems are typically designed to minimize power consumption, largely as a means of prolonging mission duration. There may be some choices about when to recharge electrically powered unmanned systems; for example, they may be able to be recharged when overall power demand is diminished. This suggests that unmanned systems are unlikely to influence a ship's peak power requirements to any meaningful degree.

Space

Unmanned systems will require space for launch and recovery, storage, maintenance, and an inventory of spare parts. Logistics and maintenance support for UASs are evolving. At present, Navy systems listed in the *Unmanned Systems Roadmap 2007–2032* are launched in highly diverse ways: using handheld bungee-launchers, runways with

---

[11] A. Estabrook, R. MacDougall, and R. Ludwig, *Unmanned Air Vehicle Impact on CVX Design*, San Diego, Calif.: Space and Naval Warfare Systems Center, Technical Document 3042, September 1998, p. 14.

[12] Department of Defense, *Unmanned Systems Roadmap 2007–2032*, Washington, D.C., 2007.

various catapults, rockets, pneumatic launchers, and others.[13] Most of the Navy's UASs are compatible with other sea-based aircraft in that rotary-blade UASs can land by hovering over a ship's deck, while fixed-wing UASs can land on the deck/runway with arresting gear. Only one variation, the RQ-15 Neptune, is recovered in open water after descending by parachute. Launch and recovery of USVs and UUVs from a surface ship may prove to be more difficult and require more auxiliary equipment onboard. For example, launch and recovery of the Remote Environmental Monitoring Unit System (REMUS) 600 is done on the surface and requires dexterity with equipment and manpower in small boats using cantilevered hoists.[14]

Ideally, future unmanned systems sharing a common domain will be capable of being launched and recovered using common spaces and equipment the way that aircraft are. Using the same launch and recovery facilities as manned systems would diminish overall space requirements, though it would also create competition for use of those facilities.

The rate of launch and recovery will affect the space requirement associated with UASs; more portals will be needed if only a limited number of platforms can be launched or recovered from a single one over a given period of time. In addition, the reliability of launch and recovery equipment will affect the desired degree of redundancy. The degree to which unmanned systems share common components will affect the mass and volume of spare parts required, as will the reliability of those components.

The fact that many unmanned systems are smaller than today's off-board systems (often manned equivalents) may not reduce space requirements, but rather result in additional unmanned systems being deployed aboard ships to make them more capable.

---

[13] Department of Defense, 2007.

[14] Daniel W. French, *Analysis of Unmanned Undersea Vehicle (UUV) Architectures and an Assessment of UUV Integration into Undersea Applications*, thesis, Monterey, Calif.: Naval Postgraduate School, September 2010, pp. 61–62.

Communications

The growing use of unmanned vehicles is likely to be associated with large increases in bandwidth requirements. Ships will be able to physically accommodate more unmanned off-board platforms than larger manned ones, so they will need to have more bandwidth to interact with those systems. Ships will need to be capable of emitting signals that can control a remote unmanned system over the din of electronic noise and receiving the muted signals of that low-power system in return. Controlling and collecting outputs from unmanned systems in data-rich combat environments will require ships to have reliable, redundant, and long-range communication capabilities.

A 1998 technical document suggested that control and planning stations for a full UAS squadron level of effort, 12 airframes with six airborne at one time, would require ten consoles for planning missions, flying airframes, and manipulating payloads.[15] Storage space for the vehicles is also a consideration. The aforementioned technical document describes one complete Predator UAS as coming in a transportation container that is 32 feet by 4.5 feet by 4 feet, on a retractable seven-inch wheel assembly, weighing about 4,000 pounds.[16]

A recent article also corroborates the utility of a ratio of four units per squadron. Four MQ-8B Fire Scout UAVs are deploying with the *Oliver Hazard Perry*–class frigate USS *Klakring*; it is reported that one operator will command two vehicles from one control station.[17] Additionally, it is reported that the ground station for the ScanEagle UAV can support the control of up to eight vehicles from two operator consoles.[18]

---

[15] Estabrook, MacDougall, and Ludwig, 1998, pp. 2 and 17.

[16] Estabrook, MacDougall, and Ludwig, 1998, p. 14.

[17] Richard Scott, "Frigate Deploys with Four Fire Scout UAVs," *Jane's Navy International online*, posted July 2, 2012.

[18] "ScanEagle, United States of America," *Naval-Technology.com*, undated.

Increasing Importance of the Electromagnetic Spectrum as a Weapon

The use of electromagnetic energy as a weapon is not new; reports of its use for this purpose date back at least to the third century B.C.[19] The use of "wireless telegraphy" for communication at sea, dating back over a century, naturally lent itself to interception, jamming, or spoofing by adversaries. Both sides used electronic warfare during World War II.[20] Moreover, since the advent of the laser in 1960, the idea of weaponizing it has been extensively explored by military personnel, scientists, and science-fiction writers.

In recent years, however, several aspects of using electromagnetic energy as a weapon have become more viable. Microwave radiation systems and laser dazzlers can now inflict incapacitating, but nonlethal, effects on personnel at a distance. These approaches enable the graduated use of force in situations in which intent may be unclear.

As unmanned platforms enable the distributed deployment of ISR sensors, the desirability of lasers (or other forms of concentrated energy) capable of blinding an adversary's sensors or damaging its thin-skinned unmanned platforms has likewise increased. Moreover, improvements in laser technology will soon enable lasers capable of inflicting substantial damage to be deployed on ships.[21] In addition to "painting" targets for targeting by projectiles, lasers can now serve as weapons in their own right.

The increasing capabilities associated with long-range targeting also make laser weapons more desirable. Since the 1960s, missiles have

---

[19] During the Roman attack on Syracuse (214–212 B.C.), Archimedes was reported to have used mirrors to focus the sun's rays on Roman ships and set them aflame. Some modern tests have corroborated the idea that such an effect could be achieved using the technology of that time, while other tests have shown contrary results. However, even if the ships could not be burned using this technology, the crews could be dazzled or temporarily blinded by the intense light.

[20] An excellent book on the subject is by R. V. Jones, *The Wizard War: British Scientific Intelligence 1939–1945*, London: Hamish Hamilton, 1978.

[21] Ronald O'Rourke, *Navy Shipboard Lasers for Surface, Air, and Missile Defense: Background and Issues for Congress*, Washington, D.C.: Congressional Research Service, R41526, April 8, 2011.

demonstrated the ability to inflict severe damage and even sink warships; improved targeting systems and increasing ranges make them a growing threat to the fleet. Preventing them from inflicting damage requires some combination of soft-kill effects (spoofing, jamming, and blinding of sensors) and hard-kill effects (damaging or destroying the missile). Both can be achieved with electromagnetic energy, either by affecting the information that missile sensors receive or by using a laser to burn through part of the missile. The abilities of electromagnetic weapons to provide an essentially unlimited magazine and to acquire new targets rapidly make them particularly valuable in the face of missile saturation attacks. While missiles can also be countered using solid weapons that emit energy to divert the missile (soft kill) or physically damage the missile (hard kill), bolstering such capabilities with electromagnetic weapons will presumably decrease the threat.

The increasingly complex nature of the battlespace and the need for coordination also make electronic warfare more important. Control of unmanned platforms as well as communication among various manned and unmanned systems are critical to combat effectiveness. The electromagnetic spectrum, like the sea itself, is a space in which no one lives but that enables vital interactions. Just as a navy ensures that its own country's use of the sea is secure while putting adversaries' maritime activities at risk, ever more electronic warfare capabilities are needed to ensure effective command, control, communications, computing, intelligence, surveillance, and reconnaissance (C4ISR) in the face of adversary attacks, and to jeopardize adversaries' C4ISR networks.

Increasing electromagnetic spectrum weapons has several implications with respect to ship design. We discuss those related to power, cooling, and space below.

Power and Cooling

Lasers and electronic warfare weapons require considerable quantities of power, particularly given that most of their energy is dissipated locally, rather than being projected by the laser beam. Surveys of laser weapon systems suggest a power efficiency of 10–30 percent, depending on the system (though this ratio may improve in the coming decades).[22] A 100-kilowatt laser would thus require something in the range of 300 kilowatts to 1 megawatt of power. A single such laser would not likely overtax existing ships' non-propulsive electrical capacity (7.5 megawatts for a *Ticonderoga* class cruiser or 9.0 megawatts for an *Arleigh Burke*–class destroyer), and some current ships might be able to support a laser with slightly more power.[23] However, substantially more powerful lasers, or multiple lasers, would exceed the capacity of a ship's electrical systems, except for the potentials of the DDG-1000 electrical system design.[24] Lasers will likely need to be powered when other ship systems, including electronic warfare and communications systems, are also operating at or near full capacity. Unless a ship contains ample power storage capacity in the form of batteries or capacitors—themselves requiring considerable space—power can be a limiting factor for inclusion of lasers in the fleet.

Lasers' low energy efficiencies imply that waste heat would thus comprise between three and nine times the laser's nominal power.[25] Other electronic systems also emit considerable quantities of heat. As a result, the cooling requirements associated with lasers and other electromagnetic weapons can require ample power and space.

Space
Lasers with 100 kilowatts or more of power take up considerable space. The 100-kilowatt free-electron lasers under consideration for shipboard use are expected to need a compartment approximately 60 feet by 12 feet by 6 feet.

Enhanced Capabilities for Long-Range Targeting
The increasing ranges at which ships can be targeted—or can target others—stem from both improved C4ISR and from the ability of cruise missiles or rail gun–launched projectiles to go ever-greater distances. As we have seen, this threat helps to drive the desirability of

---

[22] O'Rourke, 2011.

[23] O'Rourke, 2011.

[24] O'Rourke, 2011.

[25] O'Rourke, 2011.

lasers to target projectiles or damage unmanned ISR platforms as well as electronic warfare capabilities to divert missiles.

Power and Cooling

Rail guns will consume considerable energy to fire projectiles hundreds of miles. For example, a 2010 rail gun experiment entailed the use of 33 megajoules over a 10-millisecond interval, equivalent to 3.3 gigawatts.[26] This massive rate of power consumption is enabled by discharging stored energy that has been accumulated over time: a rail gun weapon system is expected to draw up to 48 megawatts of power from a ship's electrical systems, two orders of magnitude below the rate of energy discharge. However, supporting a rail gun system is more than a question of power capacity. A rail gun will likely require ship power infrastructure providing high-voltage connections around 10 kilovolts, well above what is available on typical Navy vessels.[27] Capacitors may be needed for the storage and rapid discharge of energy, while cooling will be required to counter energy dissipated within the gun. Estimates put the power efficiency of a rail gun system anywhere from 30 to 66 percent, meaning up to 34 megawatts of waste heat would need to be dissipated by ship cooling systems for a 48-megawatt rail gun.[28] These power and cooling requirements would greatly exceed those available from existing surface combatants; only a ship with an integrated power system, such as the *Zumwalt* class of destroyers (DDG-1000), could match the power requirement, and the rail gun would require power rivaling the propulsion capability.

Space

Long-range targeting systems are likely to increase space requirements. Rail guns, capacitors, and their associated cooling equipment will take up additional space. It has been estimated that a shipboard rail gun would reach a weight of 1,000 tons and take up several decks of space

---

[26] Spencer Ackerman, "Video: Navy's Mach 8 Railgun Obliterates Record," *Wired.com*, December 10, 2010.

[27] Based on previous RAND research on the rail gun's naval applications.

[28] Previous RAND research.

below the actual turret.[29] Rail gun systems in testing have a barrel length of around 40 feet, and rail gun shipboard power and control equipment are estimated to have a footprint of approximately 40 feet by 20 feet.[30] The size and weight of a rail gun system would likely necessitate its installation on a new or drastically altered platform.[31]

Moreover, when objects can be targeted at increasingly long distances, the desirability of having ample numbers of missiles or rail gun ammunition will require considerable inventory of these items. On the other hand, rail gun ammunition is smaller than conventional ammunition, and since it is non-explosive, it requires less secure storage and handling. It is unclear which effect will predominate, particularly since the amount of storage required will depend partly on the accuracy with which rail gun projectiles can strike their targets.

The Increasingly Networked Nature of the Battlespace

The ability of ships to communicate with one another, with off-board platforms, and with land bases has expanded dramatically over the past two centuries. This has been closely correlated with increasing combat capability resulting from improved coordination, as well as increasing abilities to collect, analyze, and integrate information. The ability to transmit information rapidly and effectively throughout the battlespace is critical to being able to project power. Coordination among platforms and situational awareness are also critical in defending against attack; for example, a missile targeting a ship presents a narrow profile for the targeted ship to hit, but another ship with good situational awareness can strike the missile along its longer (and less heat-resistant) side. Moreover, in an environment with high concentrations of manned and unmanned platforms from both sides, preventing fratricide while swiftly targeting enemy platforms will require extensive coordination. The result is that ships will need extensive information-

---

[29] Previous RAND research.

[30] J. Bachkosky, D. Katz, R. Rumpf, and W. Weldon, *Naval Electromagnetic (EM) Gun Technology Assessment,* Washington, D.C.: Naval Research Advisory Committee, NRAC 04-01, 2004.

[31] Bachkosky et al., 2004, p. 8.

processing and communications capabilities to perform effectively in combat, and this need has several implications.

Personnel

While IT systems involve extensive automation and much of their servicing will take place in port or ashore, the need for trained personnel to support them at sea will likely increase. IT systems can be damaged by kinetic or electronic attack, cyberattacks (including implanted "time bombs"), software bugs, and hardware defects. Software bugs and hardware defects can even be implanted deliberately, if an adversary is able to tamper with the procurement process. Personnel on the ship need to have enough knowledge to conduct repairs and maintain combat capability following attacks, in a niche type of damage control. While they can consult with external experts, their ability to do so will be attenuated by the attack itself, and they will need enough knowledge to be able to apply others' recommendations.

Communications

There will be an expansive need for bandwidth and communications capabilities to support an increasingly networked environment.

Power and Cooling

While the power and cooling requirements associated with any one system may decrease as technology advances, the explosive growth in numbers of systems may contribute to elevated power and cooling requirements.

Space

The rapidly increasing number of IT and communications systems will be offset by the decreasing size of electronics; the overall trend is unclear.

Concluding Remarks on Technological Trends

In the preceding pages, we have discussed the requirements associated with the advent of unmanned systems, increasing weaponization of the electromagnetic spectrum, longer-range targeting, and the increasingly networked battlespace. We summarize the requirements these are likely to impose in terms of power, cooling, personnel, space, and bandwidth requirements in Table 4.1.

The need for ample margins in each of these areas to accommodate future systems is apparent, even if the precise values of those margins remain unknown. Some of the margins may be diminished by a reduction in requirements associated with manned platforms, conventional naval guns, and other weapons. However, given current trends, it seems sensible to ensure that there are considerable margins in terms of power, cooling, support for personnel, space, and bandwidth.

On a final note, designers can take into account two attributes of having high margins with respect to space. The first is that additional space can be used to provide power, cooling, or bandwidth, or to support additional personnel. The second is that space is not fungible throughout the ship: extra space in one area, or distributed among several, is not equivalent to space in another location where it may be more or less useful. Warship modernizations frequently involve updating or replacing weapons and sensors. These tend to be located high in the ship's structure where extra weight has its most adverse effect on ship stability. Warships have stringent specifications for metacentric height, a measure of initial ship stability, and damaged stability (e.g., when part of the ship is flooded). Space margins left in new construction may be of little value if the desired modification violates stability specifications. Weight and stability are equally important as space margins for future modifications.

Table 4.1
Effect of Technological Trends on Ship Requirements

	Power	Cooling	Personnel	Space	Bandwidth
Unmanned systems	Little change	No change	Increase	Unclear	Increase
Electromagnetic weapons	Increase	Increase	Little change	Increase	No change
Long-range targeting	Increase	Increase	Little change	Increase	No change
Increasing Networking	Increase	Increase	More technical	Unclear	Increase

Geopolitical Context and Trends

Warships' missions are influenced by the location, nature, capabilities, and vulnerabilities of prospective adversaries. In this context, the long life spans of warships in comparison with political arrangements can pose challenges for ship design. A single battleship—the USS *Missouri*—fired naval guns against Japan in 1945 and launched Tomahawk missiles against Iraq in 1991, having lasted through the entire Cold War.[32] The aircraft carrier USS *Enterprise*, which participated in the blockade of Cuba during the 1962 Cuban Missile Crisis, finished its final deployment in 2012. Political arrangements and prospective threats can change far more frequently than warships are replaced.[33]

A review of just the past 25 years is instructive. In 1987, the effect of the novel *glasnost* and *perestroika* policies was unclear; they might have represented a ruse or a means of strengthening the Soviet Union for the next phase of the Cold War. Afghan rebels continued to fight the Soviets with U.S. support. China's economy was equivalent to a few percentage points of the U.S. economy, and its military capabilities were correspondingly limited. However, some Americans voiced disquiet about the rapid economic growth of Japan, seeing this democratic U.S. ally as a prospective challenger. Iraq, though hardly a friend of the United States, was viewed by many as a valuable bulwark against Iran; its attack on the USS *Stark* was accepted as an accident. Just three years before, during the Sarajevo Olympics, the world had applauded Yugoslavia's ability to heal its bitter wounds from World War II. In short, the conflicts that the United States has fought over the past quarter-century and the challenges that it has faced differ greatly from the ones that could have been anticipated at the beginning of the period.

---

[32] The USS *Missouri* is perhaps most famous for having been the site of the Japanese surrender at the end of World War II.

[33] See, for example, Stijn Hoorens, Jeremy Ghez, Benoit Guerin, Daniel Schweppenstedde, Tess Hellgren, Veronika Horvath, Marlon Graf, Barbara Janta, Samuel Drabble, and Svitlana Kobzar, *Europe's Societal Challenges: An Analysis of Global Societal Trends to 2030 and Their Impact on the EU,* Santa Monica, Calif.: RAND Corporation, RR-479-EC, 2013, and Gregory F. Treverton, *Making Policy in the Shadow of the Future,* Santa Monica, Calif.: RAND Corporation, OP-298-RC, 2010.

Given the rapidity of changes in the international sphere, it would be very bold to predict the precise threats and conflicts that the United States will face over the next half-century. However, the U.S. Navy can be confident that it will face diverse types of adversaries and challenges over that period; we discuss some of the probable and possible ones below.

Near-Peer Competitors

Few states have some combination of capabilities that enable them to fight in far-offshore environments and to challenge the U.S. Navy on the high seas as well as near shore. Most of these are unlikely to become wartime adversaries of the United States over the next few decades, either due to their shared interests with the United States or their democratic forms of government. Two nations stand out as possible exceptions: Russia and China. Using anti-ship missiles, long-range aircraft, long-endurance submarines, surface ships, and other systems, they could challenge the U.S. Navy's ability to operate even at long distances offshore. Moreover, both countries evince a nationalistic interest in power projection beyond their immediate home waters.

Conversely, both Russia and China face a number of challenging domestic issues that may curtail their future military capabilities. For Russia, these include population decline, political dissatisfaction, conflict in the Caucasus, and the limited strength of the economy aside from fossil fuels and mining. China struggles with an aging population, challenges in maintaining a high growth rate, environmental pressures damaging its population's health and livelihood, regionalism, and the rise of a politically savvy middle class. Endemic corruption in both societies contributes to popular discontent, as do secessionist movements and the reaction to them. The result may be that one or both of these nations may experience a diminishing set of military capabilities relative to those of the United States. The probability of conflict with the United States would also diminish as either country shifted toward a more democratic form of government or a more benign relationship with its neighbors.

We do not know whether Russia, China, or both will challenge the U.S. Navy on the high seas over the next several decades. Given this situation, over the lifetime of a warship being designed today, it is unclear whether blue-water combat capabilities will be required. Therefore, it is necessary to have an overall concept of operations that includes high-seas combat, and to design warships accordingly.

Low-End and Medium-End Threats

Nations and non-state actors will presumably continue to pose security threats for the foreseeable future. This threat includes states that can endanger the security of their neighbors, states whose hosting or sponsorship of terrorism poses a wider threat, and states too weak to impose sovereignty over their territories. It also includes capable non-state actors, who may become more capable as advanced technologies proliferate and become less costly.

Even medium-sized powers with relatively small and limited-capability surface fleets have the ability to require the U.S. Navy to engage in a diverse range of missions. Iran and North Korea, for example, both have submarines, missiles, aircraft, mines, and a number of other means of attacking U.S. warships, civilian ships, or land bases. Both have surface vessels that can be expected to swarm with suicidal intent; electronic warfare capabilities; and varying degrees of chemical, biological, radiological, and nuclear (CBRN) capabilities. The result is that to confront such powers, the U.S. Navy needs to be prepared to conduct ASW, anti-air warfare, anti-surface warfare, MCM, missile defense, electronic warfare, CBRN defense, and a number of other missions just to protect itself and key regional locations. For offensive operations, U.S. Navy ships need to be capable of launching missiles, firing guns at land targets, and supporting air operations. If a ground campaign is to be conducted, the U.S. Navy may also be providing logistical support or even supporting an amphibious assault. Throughout any conflict, the U.S. Navy will require command-and-control capabilities, assured and encrypted communications, and an array of ISR capabilities to maintain situational awareness. In short, conflict with even medium-sized powers would require a wide range of capabilities.

Both the Iranian and North Korean regimes appear fragile in a number of respects; it is unclear whether they will persist for the decades that a new warship will last. However, even if these regimes disappear early in the lifespan of a ship, the ship will still need to be designed to be able to deter or fight them for as long as they exist. Moreover, new threats may arise. Out of the world's nearly 200 nations, it would be a bold assumption to believe that, in the coming decades, no other medium-sized regional powers will come to pose a threat that the United States will need to be able to deter.

Moreover, less militarily capable states and even non-state actors may pose substantial threats requiring the U.S. Navy to engage in a wide range of missions. A striking trend of the past several decades has been the proliferation of advanced technologies and capabilities to non-state actors. Non-state actors, with varying degrees of cooperation from states, have been able to benefit from the growing availability of both advanced technology and the knowledge needed to use it. In the 2006 Israeli-Hezbollah War, Iranian-backed Hezbollah successfully targeted both an Israeli warship and (accidentally) a Cambodian civilian ship with anti-ship missiles, while also launching UASs into Israeli territory.[34] From 1984 through 2009, the Tamil Tigers—a rebel group seeking independence from Sri Lanka, largely without the support of established states—developed the world's most formidable non-state navy, complete with small submersible vessels, well-trained divers, naval mines, and raiding craft. They used these systems to sink a number of Sri Lankan vessels and to conduct substantial landings behind Sri Lankan lines. The Tamil Tigers were also able to develop and field a small air force. In South America, drug cartels have succeeded in developing semi-submersible vessels and even submarines for smuggling purposes.

The only domain in which medium-sized powers (or less-capable actors) may not be able to project power effectively is in blue-water environments. Projecting power hundreds of miles from land, in the relatively uncluttered offshore environment, remains largely the preserve of a handful of powerful states. Challenging the U.S. Navy far offshore requires either surface ships that can survive an engagement,

---

[34] The Israeli ship's missile defenses were not activated at the time, because it was not believed that Hezbollah had the capability to launch anti-ship missiles.

long-endurance submarines or aircraft, or long-range cruise missiles for targeting at sea. Even these capabilities may not be beyond what a medium-sized actor (let alone a near-peer) can achieve in the coming decades, as technologies both advance and proliferate.

Disaster Response

Natural disasters, large-scale accidents, and devastation wrought by terrorist or military attacks show no sign of abating in the 21st century. Indeed, the consequences of natural or accidental disasters may become more dire due to climate change, growing populations, and the increasing complexity of novel technologies. The political desirability of responding to these situations will also likely increase, as new technologies enable images from disasters to be better captured by witnesses and rapidly broadcast throughout the world. For these reasons, the U.S. Navy can expect to be called upon to provide humanitarian assistance and disaster relief. Within the United States, it will also have responsibilities for defense support to civil authorities and homeland defense.

Concluding Remarks on the Geopolitical Context and Trends

The proliferation of warfare capabilities to low- and medium-level threats precludes the U.S. Navy being able to assume that it can diminish its own capabilities in any particular warfare area. Moreover, the existence of prospective near-peers puts the high-seas domain—a largely protected preserve for the U.S. Navy since the Cold War—at potential risk. U.S. warship designs need to take all warfare areas and all warfare domains into consideration as possible aspects or venues of conflict.

Lessons from Past Incorporation of New Missions and Technologies

The late 19th and early 20th centuries were an era of rapid technological change for navies, as documented in *Sea Power in the Machine Age* by Bernard Brodie and *American and British Aircraft Carrier Devel-*

opment 1919–1941 by Thomas Hone, Norman Friedman, and Mark Mandeles.[35] Ships' motive power shifted from sail to steam, and the energy used to generate steam later shifted from coal to oil. Ships' hulls, which had been made of wood since ancient times, were now made of metal. Guns and ordnance grew much larger and more powerful, with ordnance increasingly comprising explosive shells rather than solid shot. Layers of armor were applied to both wooden and metallic hulls. Undersea warfare began in earnest with the advent of operationally effective mines, torpedoes, and submarines. Wireless communications enabled better coordination and improved situational awareness, while aircraft, based both at sea and on land, provided both ISR and kinetic capabilities.

Below, we explore a couple of key lessons from this era:

- Growing offensive capabilities may require new ships to survive them
- Gradual adoption of technologies and new procurement are often desirable.

Changing Offensive Capabilities May Require New Ships to Survive Them

Naval gunnery, which had varied little from the 16th through the mid-19th centuries, experienced a series of rapid changes from roughly 1850 to 1910. Within that short period, naval gunnery was reshaped by rifled barrels to improve accuracy, breech-loading to increase rates of fire, armor-piercing explosive projectiles, and improved gun construction that allowed for larger charges. It emerged that existing ships could be reconfigured to accommodate these new technologies. For example, the Italian *Duilio* and *Dandolo* warships had their 35-ton guns replaced with 100-ton guns in 1880, while new fire-control systems were installed aboard Royal Navy ships just before World War I.[36]

---

[35] Bernard Brodie, *Sea Power in the Machine Age*, Princeton, N.J.: Princeton University Press, 1941, and Thomas C. Hone, Norman Friedman, and Mark D. Mandeles, *American and British Aircraft Carrier Development, 1919–1941*, Annapolis, Md.: Naval Institute Press, 1999.

[36] Brodie, 1941, pp. 181, 198, 213, 228, 230.

However, naval authorities soon learned that withstanding other nations' new, more powerful guns required new design and construction. Less-armored or unarmored ships rapidly became vulnerable or even obsolete in the face of new weaponry; each new ship had to be designed with more armor than the one before. In many cases, simply adding armor to existing ships was not a viable option, since the armor affected displacement and balance. Big, long-range guns created incentives for larger ships that could be clad in enough armor to survive them. Moreover, topside armor became an urgent (and novel) need due to increased ranges and better fire control that enabled projectiles to hit ships from above, even before aircraft were capable of inflicting major damage against ships.[37]

Gradual Adoption and New Procurement Are Often Desirable

The advantages of steam-powered ships relative to their sail-powered rivals became evident shortly after the former were introduced in the mid–19th century. However, the technology and design standards needed time to mature. The displacements of both new steam warships and sailing ships reconfigured for steam power were typically underestimated, so they rode too low in the water. During the 1850s, when steam engines and screw propellers were rapidly improving, it made sense for the Royal Navy to transform sailing ships gradually into steam-powered vessels, to avoid overinvesting in early versions. War scares, however, induced an over-hasty rush to build.[38] The Royal Navy was generally dissatisfied with conversion of sailing vessels, because the limited space remaining aboard after steam power had been installed meant that cruising range was limited; they preferred to build new, larger vessels when timber was available. The French Navy, for which range was less of an issue, more often converted its sailing vessels.[39]

Britain had another similar experience 60 to 70 years later, as it sought to create aircraft carriers during and after the First World War. In 1917–1918, Britain removed guns from warships to turn them into

---

[37] Brodie, 1941, pp. 213–215, 232, 235, 252.

[38] Brodie, 1941, pp. 43, 57.

[39] Brodie, 1941, pp. 73–74, 76, 160.

carriers, but it found that the remaining airflow obstructions made them unsuitable platforms for supporting aircraft.[40] Britain was at the cutting edge in terms of early aircraft carriers during the interwar years, but its investment in early technologies resulted in high cost-to-capability ratios, whereas latecomers—notably the United States—often did better.[41] For example, investment in armoring carrier decks was largely negated by the advent of radar, which provided warning of incoming planes so that fighters could intercept them before they could drop bombs on the deck.[42] At the same time, having more fighters that could launch quickly became critical to protecting carriers from incoming planes.[43] The need to get large numbers of planes aloft quickly had not been obvious at the outset; whereas battleships fought attritional duels, carriers needed to strike quickly and with enough force to disable or sink an opponent. Given the limited number and scale of weapons that an aircraft could carry, this meant supporting larger numbers of aircraft than originally anticipated.

Britain's over-rapid adoption of aircraft carrier technologies had other consequences. The British tended to see the choices they made previously as both necessary (and therefore an inevitable part of their adversaries' development) and irreversible. On the other hand, subsequent developers who struggled less with the initial technology recognized additional options open to them.[44] The U.S. Navy procured carriers more slowly, experimenting with a variety of systems as the technology matured. This was beneficial in part because ship requirements expanded rapidly as aircraft grew heavier and more powerful:

---

[40] Brodie, 1941, p. 394.

[41] Hone, Friedman, and Mandeles, 1999, pp. 89–90, 167.

[42] Radar early warning and the ability to quickly launch aircraft may have been more valuable than an armored flight deck early in World War II, but the situation changed when the Japanese introduced kamikaze attacks on U.S. aircraft carriers. The kamikaze attacks resulted in major damage to the wooden flight decks of U.S carriers. In contrast, when British aircraft carriers with their armored flight decks swung to the Pacific in 1945, they experienced relatively little damage from kamikaze attacks.

[43] Hone, Friedman, and Mandeles, 1999, pp. 93, 163, 199.

[44] Hone, Friedman, and Mandeles, 1999, pp. 89–90, 105, 110.

they needed larger, stronger flight decks and elevators, more capacity to store aviation fuel, and catapults.[45] Since even small carriers needed these, this put them at a capacity-to-cost disadvantage relative to larger carriers, and the latter came to be recognized as more advantageous.

Concluding Remarks on Lessons from Past Incorporation of New Missions and Technologies

The above historical accounts highlight some of the advantages of gradual ship procurement at times of rapid technological change. In addition, they suggest that while ships can often accommodate new missions, the development of new technologies may render them vulnerable in ways that make them functionally obsolete.

Conclusions

The current geopolitical context suggests that no mission space, save those that have been entirely superseded by technological advances, can be considered irrelevant to future ship design. The substantial possibility of a near-peer challenger emerging over the lifetime of a ship, coupled with the proliferation of capabilities to lesser state and non-state actors, implies that the future fleet should anticipate all domains being contested across all warfare areas.

Millennia-old naval missions, such as surface warfare and amphibious landings, will continue alongside those of more recent vintage, such as anti-air warfare and ASW. New missions will complement the old, and new technological means of accomplishing older missions, employing unmanned systems, more-effective long-range targeting, enhanced networking of the battlespace, and increased weaponization of the electromagnetic spectrum will be fielded.

The process of introducing these capabilities, by employing novel technologies and correspondingly innovative concepts of operations, is unlikely to be smooth. Earlier transitions to steam power, metal hulls, and the employment of both undersea and aerial systems involved

---

[45] Hone, Friedman, and Mandeles, 1999, pp. 80, 82, 136, 194–195.

considerable trial and error; some ships became obsolete long before their expected service lives expired, while others were retrofitted without achieving the full capabilities associated with new systems. It will often be unclear to what degree a particular technology has matured at a given time and the extent to which its future requirements can be anticipated. Synergies among disparate systems may increase their capabilities in unexpected ways. Moreover, an adversary's use of new technologies—which may be qualitatively different—can affect ship design, missions, and obsolescence as much as the incorporation of those technologies into one's own fleet. Gradually building or retrofitting ships over time to accommodate new technologies and missions will likely save money and increase capabilities relative to the alternatives, which are to hurriedly add capabilities throughout the fleet or replace ships prematurely.

Despite the many unknown aspects of future technologies, it can be said with confidence that they will consume large amounts of power, and in doing so emit appreciable quantities of heat that need to be ameliorated through the use of cooling equipment. They will require personnel to operate them, maintain them, and often to analyze information that they have collected. These systems and those required to support them will occupy valuable space and will typically require ample bandwidth to enable coordinated operations in an increasingly networked battlespace. It is impossible to predict with any accuracy the precise requirements associated with any one system, the limiting factors that will determine how many systems will be aboard a particular ship, or the degree to which some requirements will be eliminated as antiquated systems are retired; 30 to 50 years is a long time in terms of technological change. However, ships designed with additional space for expansion and with ample margins in terms of space, power, cooling capacity, and communications capabilities will be flexible enough to adjust to a wide range of potential future systems.

CHAPTER FIVE

Roadmap for Future U.S. Navy Modularity and Flexibility Efforts

The research sought to address the following four issues:

- Categorize various modularity and flexibility options.
- Estimate the costs of the lack of modularity and flexibility in ship design.
- Understand ways to achieve adaptability when facing an uncertain future.
- Project when future opportunities will exist to adopt the concepts of modularity and flexibility.

In this chapter, we summarize the findings from the previous chapters and provide recommendations for a roadmap for modularity and flexibility in future ship designs.

Modularity and Flexibility Are Related but Different

The Navy desires adaptable ships that can be quickly and inexpensively modernized to incorporate new missions and new technologies. Modularity and flexibility are typically mentioned as concepts that contribute to adaptable ships. The U.S. Navy has used modularity concepts in several ship programs, as discussed in Appendix A. These modularity concepts typically involve self-contained modules, such as the VLS on *Arleigh Burke*–class destroyers, or modular installations, such as the LCS sea frames, and mission payload modules. Flexible infrastructures are currently incorporated into the design of *Ford*-class aircraft carriers. One additional form of modularity, the use of common modules across multiple ship classes, has not been incorporated into ship designs. Common modules directed at personnel-related functions like galleys, medical facilities, or laundry spaces could help reduce ship design and construction costs by building and testing those common modules in factory-like environments before delivery to the ship construction site. The Navy should investigate the use of this form of modularity.

Modularity typically involves the use of defined interfaces within prescribed boundaries. As long as the system fitting into the self-contained modules or used in a modular installation have those defined interfaces and can fit into the boundaries, the ship can adapt to mission or technology changes. However, adaptability is restricted if additional space or ship services are needed. Flexibility is a step above modularity in that it allows for extension of the boundaries. Flexible ships have more space and greater power, cooling, and bandwidth capabilities. The greater space and higher support services do not have to be utilized immediately, but are available when and if they are needed in the future.

Various Factors Influence the Cost of Ship Modernizations

It proved difficult to assess how modernization costs could be reduced if greater degrees of modularity and flexibility had been incorporated into the original design and construction of a ship. However, data from a recent mid-life modernization of an *Arleigh Burke*–class destroyer offer some insights into what drives modernization costs and how ship designs could reduce those costs.

A large part of installation man-hours is associated with the removal and replacement of outdated equipment or systems. Installation man-hours are driven by the amount of hot work needed to cut holes in bulkheads and other ship structures; remove and replace old foundations; and disconnect and reconnect power, cooling, and other interfaces with ship support systems. An initial analysis of a single ship mid-life upgrade suggested that the number of foundations that were removed and replaced as well as the feet of installed electrical and fiber cable were significant contributors to installation man-hours.

Improving accessibility to equipment and systems that are likely to be modernized during a ship's operational life will help reduce modernization costs. Future ship designs should examine the placement of critical spaces in the ship based not only on survivability but also on the likelihood of having major modernizations to those spaces. Future designs should consider standard foundations for certain classes of equipment, much like the interfaces that define self-contained modules or modular installations. Any new equipment would be designed to fit the standard foundation. When designing new equipment for in-service ships that have not adopted standard foundations, a trade-off analysis should be conducted to look at the costs between designing the equipment to fit the existing foundation or to remove the old foundation and replace it with a new one. Improving accessibility to equipment also involves minimizing the removal and replacement of any piping, cooling, or electrical interferences that restrict the removal of equipment. Future designs should consider the placement of ship services within a space as well as the interfaces between spaces.

Reduced modernization costs should also result from better planning and execution of the modernization work. A single manager should coordinate the development and execution of the availability period. The Navy is moving in that direction and the end result should be an improved and more-efficient process.

Little data are routinely collected on the actual costs of modernization work packages and compared with the original estimates. A concerted effort to collect, organize, and analyze data on installation man-hours and those variables that could help explain what drives those man-hours will provide additional insights into the forms of modularity and flexibility that should be adopted in future ship designs.

What the Future May Imply for Modularity and Flexibility

We examined the likely implications of both technological advances and geopolitical changes with respect to modularity and flexibility requirements. As noted earlier, four major technological trends will likely influence naval operations over the coming decades:

- The increasing use and effectiveness of off-board unmanned systems
- Growing abilities to use the electromagnetic spectrum as a weapon
- Enhanced capabilities for long-range targeting
- The increasingly networked nature of the battlespace.

By ensuring that ships have ample margins in five areas, it is possible to make them better able to accommodate these trends. These five areas are **power, cooling, support for personnel, space, and bandwidth (including concurrent weight and stability margins)**.

There is considerable geopolitical uncertainty with respect to the next few decades, both in general and with respect to the adversaries the U.S. Navy may face. These may range from near-peers to non-state actors; some of the latter may become far more capable due to the proliferation of advanced systems.

Future force structures must have a range of ships that can perform both traditional and emerging missions, including the ability to conduct traditional missions in novel ways.

Current modularity and flexibility efforts that provide the ability to change missions and technologies at low cost, and in short timeframes, must continue. These modularity and flexibility concepts should be augmented with a realization that bigger may be better. Additional volume will provide space to change interface boundaries; to increase the number of personnel; and to add power, cooling, and/or bandwidth as needed.

Where Will Future Opportunities Exist?

New ship designs provide the primary opportunity to infuse modularity and flexibility concepts. It is difficult and costly to add modularity to an existing ship design. The decision to retire the missile arm launcher cruisers earlier than their planned operational lives is one indication of the difficulty in modularizing an existing ship. The DDG-51 Flight III program is also facing the limitations of an inherited design.

In past decades, the Navy typically had a number of ship design programs underway at any point in time and new designs for certain classes of ships followed in a heel-to-toe fashion. The longer operational lives of naval ships, the ability to adapt to mission and technology changes through modularity and flexibility concepts, and budget constraints have all led to gaps in the design of new naval ships. Figure 5.1 shows the design, production, and planned operational lives for various classes of ships, based on the most recent Navy shipbuilding plan. For a ship to be a good candidate for adapting to modularity and flexibility concepts, it should just be entering the design phase or, as a minimum, be in the very early design stages. The figure shows only two near-term opportunities for new ship design: the LX(R) and the DDG-51 Flight III programs. Currently, the next new design beyond those programs could be over a decade away. Recently, the Navy announced the start of an effort to develop a new, small surface combatant to add capability beyond that provided by the LCS.

Given these near-term targets for applying modularity and flexibility in new ship designs, we now offer specific near-term and more general, overarching recommendations for future modularity and flexibility.

Recommended Future Directions for Incorporating Modularity and Flexibility

DDG-51 Flight III

Of the two near-term opportunities to embed greater levels of modularity and flexibility, the DDG-51 Flight III faces the most-severe restrictions. There is a given hull form that cannot be modified short of inserting a new mid-body section.

Even with these constraints, there are opportunities to incorporate flexible infrastructure concepts. The internal spaces can and will be adjusted to provide greater power and cooling capabilities. When spaces are modified, the new decks and bulkheads could adopt the

Figure 5.1
Potential Targets for Modular/Flexible Ship Designs

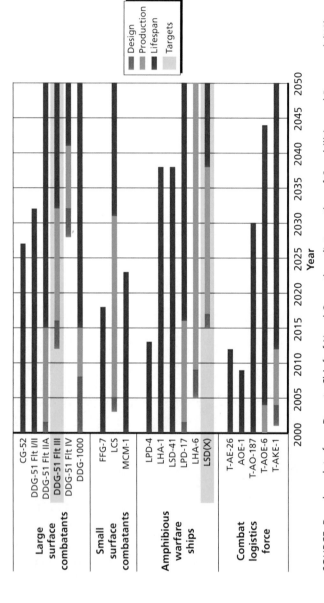

SOURCE: Based on data from Deputy Chief of Naval Operations (Integration of Capabilities and Resources) (N8), *Report to Congress on the Annual Long-Range Plan for Construction of Naval Vessels for FY2014, Washington, D.C.: Office of the Chief of Naval Operations, May 2013.*
RAND RR696-5.1

same track and interface concepts used in the *Ford*-class carrier design. A second opportunity is to design and build common hotel-related modules that are useable across multiple ship classes. NAVSEA, the shipbuilder, and the appropriate equipment contractors could work together to define a needed space with required interfaces for such applications as a medical facility, a laundry room, or berthing compartments. The shipbuilder would build the space, and a potentially separate organization could be designing and building the module to insert into the space. The common module(s) could be used in the design of other classes of ships, for example, in the design of the LX(R).

LX(R)

The LX(R) offers several opportunities for incorporating more modularity and flexibility into the design of the platform. A design with greater volume and power than the current class will provide room for potential future growth as well as a less-dense ship to operate and maintain. The *San Antonio*–class hull form has been examined as a candidate for the new LSD. The LSD size has grown over the past four classes, as shown in Table 5.1. The length of the *Harpers Ferry* class does not differ much from that of the *San Antonio*. However, the latter has a full-load displacement 50-percent greater than that of the *Harpers Ferry* class, and changing the length and/or beam of an existing ship design would require a significant redesign effort.

We offer two recommendations for the new LX(R) design. First, it should continue to incorporate self-contained payload modules and modular installations. Large, open mission bays with connections for standard modules should be considered during the Analysis of Alternatives. The Danish Navy's *Absalon*-class flexible support ships have such a "flex deck" mission bay, with the ability to host various mission capabilities that can fit either inside standard ISO shipping containers or expandable shelters compliant with ISO standards. An example the Danish Navy has developed is an embarkable medical treatment facility that can perform ten surgical operations in general anesthesia and 30 to 40 emergency treatments per day. These modules enable the ship to perform multiple missions by quickly changing the modules that are required. The ship retains the inherent ability to embark payloads of up

Table 5.1
Characteristics of Different Classes of LSDs

Class	Length (feet)	Full Load Displacement (tons)	Marine Detachment
Thomaston	510	11,300	330
Anchorage	553	14,000	330
Whidbey Island	610	16,000	400
Harpers Ferry	610	16,500	504
San Antonio	684	25,000	N/A

to 55 vehicles or 7 main battle tanks in the same space. Second, we recommend that the new LX(R) design also consider flexible infrastructures in various spaces and the development and use of hotel-related modules that can be used across multiple ship classes.

Overarching Recommendations

In addition to the specific program-related recommendations for the future of modularity and flexibility in naval ship designs, we make the following overarching recommendations:

- The Navy should continue to encourage and develop the concepts of modularity and flexibility, but do so in a more focused, coordinated fashion. The various program executive officers should coordinate a development of Navy policy on the use of modularity and flexibility in new warship designs. The new manager will coordinate repair and modernization work packages and could play a role in the development of the future concepts of modularity and flexibility that apply across different classes of ships. This could include having input into the design of common modules for hotel-related functions and the use of common foundations for equipment. This organization would work with the shipbuilders, the major combat and weapon system designers and manufacturers, and the vendors that provide hotel-related systems during any design efforts.

- The Navy should initiate efforts to gain a better understanding of the modernization costs of in-service ships, the factors that drive those costs, and how the costs could be reduced if greater levels of modularity and flexibility were incorporated during the design of a ship.

There is mixed support for greater levels of modularity and flexibility in ship designs to provide ships that can adapt to future missions and technologies in a cost-effective manner. The Navy must continue to identify and realize opportunities to provide an affordable and adaptable fleet.

Past Efforts Toward Adaptability

The idea of incorporating modularity and flexibility into the design of surface combatant ships is not a new one. The U.S. Navy has been studying modularity for over three decades in a series of evolving programs, motivated mainly by rising ship costs and shrinking budgets. These programs have looked at all aspects of modularity: whether modularity is a good idea, how to implement modularity into ship design, and the potential drawbacks and benefits (see Table A.1). Additionally, many of the programs have envisioned a modular concept ship that incorporates modularity in all aspects of its design. However, despite the decades of work devoted to the study of modularity, the concepts and conclusions reached by these efforts appear to have had mixed impacts on U.S. surface combatant design.

Foreign navies have also used various forms of modularity. Table A.2 describes two foreign navy modularity programs, namely the German MEKO and Danish StanFlex programs.

U.S. Modularity Efforts

SEAMOD—NAVSEA, 1975–1978

Study of modularity as applied to U.S. Navy ships began in 1975 with the Sea Systems Modification and Modernization by Modularity (SEAMOD) program. SEAMOD, within NAVSEA, was established as a response to apparent issues with the state of ship acquisition at the time. Growing complexity in surface combatant design, construction, outfitting, and modernization seemed to be producing highly

Table A.1
U.S. Modularity Efforts

Year(s)	Program	Description	Result	Impact
1972–1978	SEAMOD	Study of the concept of designing a ship to receive a modularized combat system to lower life-cycle costs, introduce ships and weapon systems faster, and maximize fleet effectiveness.	Analysis of concept showed that "advantages outweigh its penalties," and that a modular ship would be over 100% more effective than a conventional ship.	First study examining a modular shipbuilding concept, work continued in SSES follow-on program
1980–1985	SSES	Follow-on of SEAMOD modular ship concept, applying work to DDG-51. Explored concept of a multi-mission modular ship to replace frigates, destroyers, and cruisers.	Developed VPS concept, showed modularity could simplify ship construction and allow modification of weapon loadouts without major ship alterations.	Led to installation of A/B modules on DDG-51 VPS concept realized by Blohm & Voss MEKO
1992–2003	ATC	Focused on reducing ship costs through commonality, standardization and modularity, anticipating future budget constraints, and analyzing several case studies.	Showed implementation of commonality and modularity as an effective method of cost reduction in case studies.	Program formed TOSA team, influenced many future Navy commonality and modularity efforts
1994–2004	OSJTF	DoD-wide task force focused on decreasing costs and increasing interoperability and modularity in future combat systems.	Developed MOSA guide and rating system for compliance.	MOSA guide and design approaches is the design standard for future combat systems
1998–2003	TOSA	ATC program team to develop physical and functional interface standards forming the "building blocks" of a ship Open Systems Architecture.	Developed framework for developing interface standards for modular payload ships.	TOSA team continued in the LCS MSSIT, developing interfaces between LCS sea frame and LCS mission packages

Table A.1—Continued

Year(s)	Program	Description	Result	Impact
2003	OACE	Effort to increase combat system software modularity, motivated by a need for open architecture standards for combat system software.	Selected open standards for the following: physical media, networks and protocols, operating systems, middleware, and languages.	Key principles incorporated into DoD IT Standards Registry (DISR), influenced open architecture in Aegis and Ship Self Defense System software
2003–Present	AIMS	Evolution of ATC program and TOSA, motivation to promote increased ship system modularity and interface standards—facilitating technology refresh and insertion, decreasing life-cycle costs, and increasing mission readiness and mission flexibility.	Formed Modular Adaptable Ship concept: implementation of MOSA at the total system/ship level.	Supports future Navy MAS plans, as well as acquisition of consumer technology allowing for greater modularity and mission flexibility
2003–Present	LCS MSSIT	Management of module and ship systems development and integration for LCS.	Oversees modular interface development and integration for LCS sea frame and mission packages.	Ongoing, TBD

Table A.2
Foreign Modularity Efforts

Year(s)	Program	Description	Result	Impact
1981+	MEKO	Modular ship construction in an effort to reduce construction costs, an evolution of the VPS concept: platform + payloads	MEKO ships were successfully built and claimed reduction in life-cycle costs	MEKO ships sold to and in use by many world navies
1986+	STANFLEX	Shipbuilding effort to use modularity concepts to achieve mission flexibility—replacing several classes of ship with a single multi-role class.	Successful implementation of STANFLEX platforms and mission modules: a fleet of standard ships that can be easily upgraded and serviced and quickly repurposed for missions as needed	Influenced development of modular design in LCS

integrated ships with long acquisition cycles.[1] This meant ships could no longer be updated effectively, due to both the length of the acquisition cycle being outpaced by changes in technology and the work involved in replacing integrated systems. SEAMOD hoped to relieve these issues by introducing modularity: functionally separating ship systems into modules, allowing them to be easily installed, repaired, and replaced—a philosophy known as "Design for Change." It was also expected that the SEAMOD modular ship concept would allow for the asynchronous design and production of systems, shortening the ship development cycle.[2]

The SEAMOD concept envisioned the development of surface combatant ships as austere "platforms" with little functionality on their own. Each platform was conceived as supporting a payload—a group of pre-packaged modules providing a set of capabilities to a ship. Several benefits of this approach were anticipated:

---

[1]   C. E. Lawson "SEAMOD—A New Way to Design, Construct, and Modernize Navy Combatant Ships," *Naval Engineers Journal*, Vol. 90, No. 1, 1978.

[2]   See J. V. Jolliff, "Modular Ship Design Concepts," *Naval Engineers Journal*, Vol. 86, No. 5, October 1974.

- Modules could be individually developed to a set of interface standards in parallel with platform development.
- Modules could be rapidly installed, repaired, or upgraded.
- Ships could be tailored to provide flexibility, allowing the ship to be adapted to specific situations.[3]

The project determined that the SEAMOD concept necessitated the establishment of a dedicated shore facility for the installation, integration, and upgrade of modular systems. This Module Installation Facility, or MIF, was seen as an evolution of the typical shipyard, with features supporting the change out of combat system modules for repair or mission change.

In addition to developing the concept of a modular payload ship, the SEAMOD project worked toward its implementation. The project developed requirements for the design of a modular ship, including functional distinctions between ship systems and definitions of power, water, and signal interfaces between platform and payload. Furthermore, the project studied the overall feasibility of its work using the USS *Spruance* (DD-963) and USS *Oliver Hazard Perry* (FFG-7) as templates.[4]

The study's findings about the benefits of the SEAMOD concept were generally positive. First, it was concluded that implementation of the SEAMOD concept was feasible: It would be possible to construct such a modular ship while maintaining seaworthiness. Second, such an implementation was not seen to cause any major deficiencies in effectiveness or availability of ship services. In general, the program found that the development of SEAMOD ships would save time in most aspects of ship development and lifetime support: off-line time, shipyard time, and installation time for new technology, as well as cost savings in most areas, save for payload acquisition costs. Additionally, overall ship effectiveness was predicted to increase significantly, as SEAMOD ships could be kept more up-to-date, with shorter time lags between technology introduction and integration. From a fleet-wide

---

[3] Jolliff, 1974.

[4] See Lawson, 1978.

perspective, the study found that total implementation of SEAMOD would enable a reduction in the number of surface ships and payloads while maintaining force-level requirements.[5]

As the first official program studying modularity and flexibility for surface combatant ships, SEAMOD inspired future efforts, such as the Ship Systems Engineering Standards (SSES) program.[6] However, since SEAMOD, the adoption of modular design concepts into actual surface ship programs has been limited. In fact, nearly two decades after SEAMOD, a paper entitled "Is It SEAMOD Time?" found that little progress had been made toward modularity. Additionally, the paper noted a resistance to change in the naval engineering community, and expressed a need to reconsider the implementation of the SEAMOD concept.[7]

SSES—NAVSEA, 1980–1985

The SSES program, also within NAVSEA, was an effort that continued the study of modularity and flexibility that SEAMOD began. The program's motivations were similar to those of SEAMOD, with the SSES Program Management Plan stating, "The current methods of Naval shipbuilding and ship acquisition are increasingly being demonstrated to be overly expensive and inadequate to meet the Navy's needs in the present era of rapidly changing enemy threats and Navy mission scenarios." Additionally, the plan noted that the benefits of series production are typically lost in the traditional shipbuilding model, with the long development cycle and continual changes during the construction of a ship class. SSES work focused on creating interface design standards for the development of modular ships. The program's vision

---

[5]   See Lawson, Charles E., "SEAMOD—A New Way to Design, Construct, Modernize, and Convert U.S. Navy Combatant Ships," *14th Annual Technical Symposium*, Association of Scientists and Engineers, 1977..

[6]   See Raymond T. Marcantonio, E. Gregory Sanford, David S. Tillman, and Andrew S. Levine, "Addressing the Design Challenges of Open System Architecture Systems on U.S. Navy Ships—Building Out of the Box," MAST 2007 Conference, 2007.

[7]   See Wade A. Webster and William D. Tootle, "Is It SEAMOD Time?" *28th Annual Technical Symposium*, Association of Scientists and Engineers, 1991.

of a highly modular ship was evolved from SEAMOD into the VPS concept.

It was noted that in traditionally designed ships, combat systems are tightly integrated with the ship platform, with many systems using unique interfaces. By specifying interface standards, the program hoped to induce a functional separation between a ship platform and combat systems in future ship designs. The VPS concept was developed as a potential ship that would make use of all of the interface standards and modular architecture features developed by the SSES program. The concept envisioned a ship platform with empty zones that would support the installation of modular combat systems by providing appropriate interfaces. The nature of these interfaces and the general zonal architecture of a VPS ship were laid out by SSES.

The VPS concept formalized the idea that modularity could not only ease the installation and modernization of ship systems, but could also allow for alternate systems to be installed. The SSES vision of a ship with a "variable payload" is of a platform that can be adapted, converted, or tailored by installing a different suite of modules.[8]

The overarching goal of the SSES program was to implement the VPS concept of a modular ship. To reach this goal, the program's objectives were to develop Ship System Engineering Standards, a complete set of standards to be used in the development of a new generation of modular surface combatants, as well as the development of generic VPS designs that adhered to these standards.

As the program progressed, work on SSES was directed toward creating modular interface standards for the next destroyer class, that of the USS *Arleigh Burke* (DDG-51). A framework of interface standards was developed for combat system elements and ship zones, but the only modular elements that ended up in the final design were those for the "A/B" weapons zones supporting the VLS.[9]

---

[8] See J. Vasilakos, R. Marcantonio, and S. Garver, "A Guide for the Design of Modular Zones on US Navy Surface Combatants," Naval Sea Systems Command, SER-4/05T, January 25, 2011.

[9] See S. Garver, R. Marcantonio, and P. Sims, "Modular Adaptable Ship (MAS) Total Ship Design Guide for Surface Combatants," Naval Sea Systems Command, SER 9/05T, February 7, 2011.

Some in the ship design community have criticized the DDG-51 design for not applying SSES modularity principles more widely; suggesting that expensive redesign in later flights of the class could have been avoided if the SSES interfaces were put in place initially. However, only the preliminary SSES VLS module was available when the contract design was completed.

ATC—NAVSEA, 1992–2003

The Affordability Through Commonality (ATC) program, beginning in 1992 within NAVSEA, continued the U.S. Navy's study of modular ship design. The initiative was motivated in part by rising acquisition and life-cycle costs for surface combatants—an "affordability crisis" resulting from rising costs coupled with downward pressure on the defense budget because of the end of the Cold War. Other factors prompting the study of commonality were a shrinking maritime industrial base and the need for operational flexibility in facing uncertain future threats. The primary objective of the ATC program was to reduce costs by improving the processes involved in every step of the ship development cycle. As the name suggests, there was a particular focus on commonality. Commonality was defined by ATC as a "synergistic combination of three pillars":

- equipment modularization
- increased equipment standardization
- process simplification.[10]

The program proposed a move away from specialized ship classes with their own unique system designs and processes. To achieve this, the ATC approach was outlined, including strategies and policies for implementation. It envisioned a Navy with fewer standard system designs used by more ships, equipment procured for fleet-wide use, and increased parallel development.

---

[10] I. M. Cecere, J. Abbot, M. L. Bosworth, and T. J. Valsi, "Commonality-Based Naval Ship Design, Production and Support," Naval Sea Systems Command Dahlgren, November 1993.

In addition, the ATC effort studied the importance of decisions made in the first stages of a new ship design. It was found that while the initial development (concept and preliminary design) of a ship consists of only 2 percent of total life costs, choices made early "lock in" the other 98 percent of costs. Thus, the program encouraged ship designers to have an increased awareness of life-cycle costs, rather than focusing on decreasing design or construction costs.

Significant barriers to increased commonality and standardization were identified, mostly having to do with the pace of the ship development process. The length of time between major shipbuilding programs meant that common equipment or procedures might not be available or effective for re-use between ship classes—whether due to obsolescence, manufacturer turnover, or other factors.

Cost-benefit analyses of the ATC approach showed decreases in procurement cost, installation cost, and labor time across a number of case studies. These reductions were found to be the result of the lower cost of COTS components, the ability to purchase common components across ship types, and the enforcement of interface standards allowing for parallel assembly of systems.

ATC began the Total Ship Open Systems Architecture (TOSA) effort in 1998.

OSJTF—OSD, 1994–2004

Parallel to Naval efforts toward modularity, the Open Systems Joint Task Force (OSJTF) was formed in 1994 by a directive from the Under Secretary of Defense for Acquisition and Technology. The OSJTF was a DoD-wide effort to change defense acquisition to make use of a Modular Open Systems Approach (MOSA) wherever possible. As defined by OSJTF, MOSA is "An integrated business and technical strategy that employs a modular design and, where appropriate, defines key interfaces using widely supported, consensus-based standards that are published and maintained by a recognized industry standards organization."[11] Similar to the SSES program, MOSA encourages an approach to the acquisition and development of combat systems governed by modular design and clearly defined interface standards.

MOSA was developed as a group of principles that could be followed by system engineers and program directors to

- establish enabling environment
- employ modular design
- designate key interfaces
- select open standards
- certify conformance.

The methods to implement these principles are described more completely by the *Program Manager's Guide: A Modular Open Systems Approach (MOSA) to Acquisition* developed by the task force.[12] As with the ATC program, the OSJTF emphasized that MOSA should be addressed early in the planning of an acquisition program, because initial design decisions were seen to have a large effect on the later openness of a system.[13]

The task force expected a number of benefits from the implementation of MOSA in acquisition programs, and shared some of the motivations for Naval modularity efforts: decreased life-cycle costs, increased interoperability, and improved upgradeability.[14]

TOSA—NAVSEA, 1998–2003

In 1998, the Total Ship Open Systems Architecture (TOSA) Integrated Product Team was formed out of the ATC program office. As before, this effort was motivated by pressure to achieve an affordable and effec-

---

[11] See Open Systems Joint Task Force, *Program Manager's Guide: A Modular Open Systems Approach (MOSA) to Acquisition*, U.S. Department of Defense, Version 2.0, September 2004.

[12] Open Systems Joint Task Force, 2004.

[13] See C. H. Azani and K. Flowers, "Integration, Business and Engineering Strategy Through Modular Open Systems Approach," *Defense AT&L Magazine*, January–February 2005.

[14] See Open Systems Joint Task Force, 2004.

tive adaptable fleet by lowering total life-cycle costs and decreasing the complexity and time involved in the ship development cycle. Because of changes in Navy acquisition policy, TOSA focused on the application of Open Systems Architecture (OSA) concepts to Navy ship development, moving away from the emphasis on standard hardware and COTS solutions.

The TOSA team noted that traditionally, ship classes are designed independently and uniquely, essentially "point solutions" to specific mission requirements that exist at particular times. However, because of the ever-increasing rate of change of technology and downward pressure on the Navy budget, a move away from the traditional way of doing things was desired. Thus, the OSA concept was developed. The TOSA effort defined an OSA as

> sufficient open standards for interfaces, services, and supporting formats that enable properly engineered elements to be used across a wide range of platforms with minimal changes. . . the equipment can be replaced by different products or new technologies with like function and capacity without requiring changes to the system's support services, control functions, or structure, and can operate successfully when the new equipment is installed.[15]

The goal of the project was to encourage the use of these OSA principles in ship design, by developing processes and open standards that ships and systems could be designed to.[16]

The TOSA team created a process by which OSAs could be designed, as well as actually developing OSA concepts for the combat information center (CIC) and some HM&E systems. In addition, TOSA engaged in technology management for the DD-21 *Zumwalt* and LCS ship programs. The TOSA process was created as a guide to identify candidate systems and interfaces for developing OSA:

* requirements

---

[15] R. Devries, K. T. Tompkins, and J. Vasilakos, "Total Ship Open Systems Architecture," *ASNE Naval Engineers Journal*, Vol. 112, No. 4, July 2000.

[16] See Devries, Tompkins, and Vasilakos, 2000.

- reference models
- architectures
- interfaces
- products

The team encouraged the use of total ownership cost as a metric for determining the value of using OSA in ship and system design.

Similar to previous programs, the TOSA Integrated Product Team also developed its own vision of a ship using open systems, open zones, and open interfaces: the Adaptable Ship. As with the VPS concept of the SSES program, the Adaptable Ship employed zonal architecture, ship spaces providing open interfaces allowing for easy installation and replacement of systems. The Adaptable Ship provided zones beyond just combat systems, like the VPS concept. Zones were outlined for ordnance, machinery, C4I, CIC, and other types of systems.

The follow-on program, Architectures, Interfaces, and Modular Systems (AIMS), began in 2003, evolving from the work of TOSA.

OACE—NAVSEA, 2003

The Open Architecture initiative was created under the Assistant Secretary of the Navy for Research, Development, and Acquisition to impel the development of open architecture systems within the Navy. As part of the overall open architecture strategy, the establishment of Open Architecture Computing Environments (OACEs) was called for. Similar to the work done in the SEAMOD program, OACE specifically addressed the use of open and adaptable software architecture. OACE was motivated by many of the same issues as hardware/physical modularity programs: life-cycle cost, difficulty involved in technology refresh, and the time involved in upgrading and maintenance.

In fact, OACE has many features in common with contemporary concepts of physical modularity. The OACE Design Guidance defines and enumerates the characteristics of an ideal open system, including the adherence to public standards (for interfaces), the use of common products supported by stable vendors, portability, and simple scalability and upgradeability.[17]

The interface standards for data and software developed by OACE are part of the MAS concept developed by the AIMS program.

AIMS—NAVSEA, 2003–Present

The Architectures, Interfaces, and Modular Systems (AIMS) program began in 2003 as an evolution of the work of ATC/TOSA. The objectives of AIMS are similar to those of previous programs, including achieving a reduction in life-cycle costs, increasing effectiveness through technology refresh, and making ships more adaptable.

As with SEAMOD and SSES, the AIMS program's vision was to engineer a ship that made full use of adaptable design principles; this concept was called the Modular Adaptable Ship (MAS). The *Modular Adaptable Ship (MAS) Total Ship Design Guide for Surface Combatants* notes the limited success of previous modularity efforts:

> Early attempts to incorporate modularity into combatants were made after the hull form was selected and weight, space, and material budgets were set. The result was the rejection of many cost saving modularity features . . . the U.S. Navy has recognized the need to incorporate modularity features from the very earliest phases of the ship design process.[18]

The MAS concept was envisioned by the AIMS program as the ideal implementation of a fully modular and flexible design. It included modular zones with open, standard interfaces for most functional areas of the ship: C4I, weapons, machinery, sensors, etc.

The work of the AIMS program directly led to the development of LCS with open architecture systems and modular mission packages decoupled from its sea frame. It also contributed modular and adaptable features to other ship programs, such as the *Ticonderoga*, *Arleigh Burke*, and *Ford* classes.

---

[17] Naval Surface Warfare Center Dahlgren Division, *Open Architecture (OA) Computing Environment Design Guidance*, Version 1.0, August 23, 2004.

[18] See Karvar et al., 2011.

LCS MSSIT—NAVSEA, 2003–Present

The LCS is the first U.S. ship design to implement the modular repurposing ability envisioned by the SSES program's VPS concept (in fact, this idea was mentioned as early as the SEAMOD program). The basic LCS is made up of the hull, HM&E systems, and core systems providing basic functionality and seaworthiness. The sea frame features mission zones: spaces supporting the installation and replacement of modular mission packages. The LCS can be repurposed to carry out three missions through the installation of these mission packages: ASW, SUW, and MCM.

The LCS features only these three mission packages currently, but the modular nature of the sea frame and mission zones might allow for the LCS to take on new missions through the development of future mission packages. Mission packages themselves are made up of a group of modules—boxed systems (in commercial shipping containers) with external interfaces that, combined, provide the functionality needed to perform all the tasks required of a mission.

To develop the interfaces between the sea frame and mission packages, the LCS Mission System and Ship Integration Team (MSSIT) was formed (mostly out of the TOSA Integrated Product Team). The MSSIT has oversight over the development of modules and ship systems, attempting to ensure that mission packages are able to integrate into the sea frame, as well as that these packages can be installed and changed easily.[19]

Other Navy Ship Programs

While the programs discussed above represent programs with a specific focus on modularity or flexibility, there have been other Navy programs that have exhibited characteristics of modularity or flexibility. The Navy's DDG-1000 program is an example.

Electronic modular enclosures (EMEs) have been used on DDG-1000 for various support equipment, including sonar, radar, communication, and data centers. Interface is at the room level, where the ship

---

[19] See P. Cheung, A. Levine, R. Marcantonio, and J. Vasilakos, "Standard Process for the Design of Modular Spaces," *ETS 2010*, Paper No. SNAME-047-2012, June 28, 2012.

provides water, power, cooling, and data attaching at I/O panels. The use of EMEs allows for room installation and testing of equipment off-ship and provides a contained and protective environment for the electronics. The EMEs also provide shock qualification (also at rack level), security, and remote monitoring that includes fire safety.

Foreign Modularity Efforts

MEKO—Blohm + Voss GmbH, 1970s–Present

During the late 1970s, in an effort to develop inexpensive, customizable ships for sale to various world navies, German shipbuilding firm Blohm + Voss developed the MEKO concept. MEKO, from the German *mehrzweck-kombination* or "multipurpose-combination," is a family of small surface combatant ships (displacement 1,000–4,000 tons) that make extensive use of modular design principles. The modular design of MEKO has several purported benefits:

- Lower construction cost from building large runs of identical hulls and systems, to take advantage of economies of scale.
- Ease of design and construction with parallel development of platforms and modules.
- Cheaper and simpler customization, conversion, and modernization by implementing a SEAMOD-like functional decoupling of platform and payload.[20]

The MEKO family of ships was the first physical realization of the full-ship modularity concept developed by SEAMOD. To date, Blohm + Voss has built 60 MEKO ships for 11 world navies. They have reported reductions in both construction time and cost over traditionally configured ships.

---

[20] See Garver, Marcantonio, and Sims, 2011.

Cellularity—UK Royal Navy, 1985

In 1985, naval architect P. J. Gates presented a paper on the cellularity concept. It was intended to reduce problems installing and removing electronic equipment and reduce the requirement for difficult structural work during conversions and modifications.

Cellularity regulated the size of ship spaces such that the minimum size of transport passageways, hatchways, and doors could not be smaller than the maximum size of electronic cabinets and compartments that were to be installed.

STANFLEX—Royal Danish Navy, 1985–Present

In 1985, the Royal Danish Navy had an operational need to replace 22 small surface warships. Because of budget constraints, one-for-one replacement was unachievable. The Standard Flex 300 (STANFLEX) project was created as a way to reduce operating costs while still maintaining the fleet's operational capabilities.

The STANFLEX concept envisaged a ship platform configured to accept interchangeable combinations of modules depending on mission requirements, as well as a fleet composed entirely of these platforms. Feasibility studies indicated that only 16 multi-role STANFLEX ships would be required to accomplish the missions previously requiring 22 traditional (non-modular) ships.[21]

Each STANFLEX ship has four modular payload bays, allowing for the following mission variants:

- surface attack
- ASW
- MCM
- minelaying
- patrol/surveillance
- pollution control.

---

[21] See Robert O. Work, *Naval Transformation and the Littoral Combat Ship,* Washington, D.C.: Center for Strategic and Budgetary Assessments, 2004.

These modular mission packages can reportedly be installed and tested in a few hours, allowing for rapid transition of ships. As a caveat, the Royal Danish Navy has noted that role-switching necessitates additional specialist training for the crew. Another benefit of the STAN-FLEX modular design is that mission modules can be reused by other vessels when a ship or class is removed from service.[22]

---

[22] See Scott, 2012.

Modularity and the DDG-51 Program

The original concept of a VPS was a product of the mid-1980s modularity programs. The DDG-51 program was still completing early stages of program acquisition[1] and was an opportunity for incorporation of modularity concepts being studied at the time. In this appendix, we examine the DDG-51 program as a case study for assessing modularity in surface ship design. This case study was performed as a macro-assessment of several key issues uncovered during the course of this research.

Assessing Modularity for the DDG-51 Program

Arleigh Burke–class ships are multi-mission guided missile destroyers intended to be operated independently as well as with other warships, such as aircraft carriers, in a multi-threat environment (air, surface, and underwater threats). In the mid-1980s, the DDG-51 was in the early stages of design and acquisition. Contract design was in the final stages and detail design would commence with the award of the lead ship contract to Bath Iron Works on April 2, 1985. The first ship in the class, the USS *Arleigh Burke* (DDG-51), was commissioned in July of 1991. Almost three ships per year have been commissioned during the

---

[1] The Navy completed the DDG-51 concept design in December 1980, completed the preliminary design in March 1983, and awarded the lead ship detail design and construction contract to Bath Iron Works in April 1985 (Department of Defense, *Selected Acquisition Report: DDG-51*, Washington, D.C., December 13, 2011b).

past 20 years. The class has been acquired in flights, commonly designated Flight I, Flight II, Flight IIA, and Flight III.[2] Flight II, introduced in fiscal year 1992, incorporated combat system improvements to the SPY radar system and the Standard missile, as well as upgrades to active electronic countermeasures and communications. Flight IIA, introduced in 1994 to the DDG-79 hull and follow-on destroyers, added a helicopter hangar with the ability to support two helicopters. Flight III refers to the current DDG-51 design effort.

Figure B.1 provides an acquisition overview for these ships. The bars on the chart reflect the time between the following major events: contract award, lay keel, launch, deliver, and commission. Note that the DDG-51 design has changed over time with a large incremental design change occurring at Flight IIA.

Our assessment of modularity and the DDG-51 program should be considered a macro-assessment of significant issues. We focus our assessment on three such issues (which also provide the organizational framework for this appendix):

- What were the previous flexibility and modularity initiatives and plan for the DDG-51 class? How were the plans executed?
- Did the lack of a total modular adaptable ship design keep important capabilities from entering the fleet?
- What do modernization efforts cost and how are those costs allocated between design, production labor, and materiel?

Flexibility/Modularity Initiatives for the *Arleigh Burke* Class

Our assessment begins with the review of two modularity initiatives and their interaction with the DDG-51 program. These are the SSES program and the Aegis combat system (ACS).

---

[2] Studies on upgrades for Flight II and IIA (originally called Flight III) were conducted during the design of the Flight I ships. These upgrade studies were actually completed before the launch of the first ship in the class.

Figure B.1
Overview of DDG-51 Acquisition

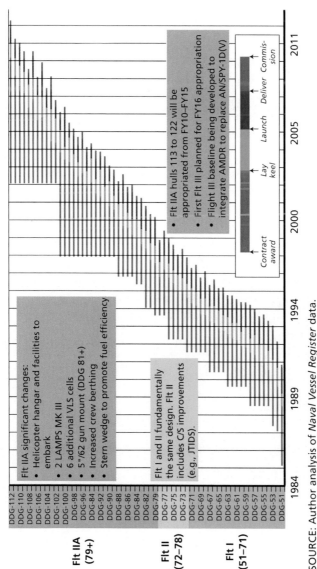

SOURCE: Author analysis of *Naval Vessel Register* data.
NOTE: *C/S* = combat system; JTIDS = Joint Tactical Information Distribution System; AMDR = Air and Missile Defense Radar.

RAND *RR696-B.1*

DDG-51 and the SSES Program

We have previously introduced and discussed several of the formal Navy programs seeking to develop and explore the notion of modular combat systems. The program with the closest ties (both conceptually and relative to the early design) to the DDG-51 was the SSES program. This program expanded upon the SEAMOD program to develop engineering standards and interface specifications with the purpose of decoupling ship weapon systems from the ship platform design.

An important objective of the SSES program was determining ways to inject modularity initiatives into Navy shipbuilding plans. The DDG-51 was in the early stages of design and a seemingly natural fit to assess for incorporation of SSES principles. In January 1983, the SSES program produced the DDG-51 Variable Payload Design Impact Study,[3] which assessed the impacts of implementing SSES on the DDG-51 preliminary design baseline. The core assessment determined the extent of SSES features in the current design and the necessary design changes for and impact of a full SSES implementation.

It is important to note that in May 1983 only preliminary SSES guidelines were developed and a full set of standards had not been approved by the Navy. Thus, partial implementation became the design target. The initial and best-suited candidates were the A-size (32-cell forward VLS) and B-size (64-cell aft VLS) module stations and support zones. A fully developed SSES design would consider the "most demanding" combat suites that a weapon zone may be required to support. Herein lies the challenge for any modular design—a design commitment to primary and support services standards is required that may not be required for the system of today, but provides for an installation or modernization effort in the uncertain future.

Table B.1 provides a comparison that helps illustrate the nature of this challenge. The first column provides the modular design spec-

---

[3] Our understanding of this report comes primarily from P. Beurman, K. Lew, and J. S. Webster, *Ship System Engineering Standards Implemented on DDG 51 Lead Ship Retrospective*, Arlington, Va.: Gibbs & Cox, Inc., August 1996. The original report is referenced as: Department of the Navy, *DDG51 (VP) Design Impact Study*, prepared by John J. McMullen Associates, Arlington, Va., Contract No. N00024-82-C-5344, TI-0001-007, TI-0002-0011, January 15, 1983.

Table B.1
SSES for Connected Loads Were Partially Implemented for the B-Size Weapon Zone (64-Cell Aft VLS)

Support Service	Modular Design Specifications (considering "most demanding" potential modules)	Modular Design Specifications (considering "VLS-like" replacements)	DDG-51 Requirements per Contract Documents[a]
Electrical			
60 Hz/30/440V	550 kw	182 kw	550 kw
60 Hz/30/115V	0 kw	0 kw	0 kw
60 Hz/10/115V—Lighting	12 kw	7 kw	12 kw
400 Hz/30/440V	75 kw	45 kw	75 kw
Fluids			
Firemain	4,000 gpm	2,740 gpm	1,750 gpm
WDCM	40 gpm	req'd	req'd
Drainage	2,000 gpm	1,370 gpm	2,660 gpm
Chilled Water	112 gpm	49 gpm	108 gpm
Potable Water	15 gpm	0 gpm	0 gpm
HP Compressed Air	160 scfm	0 gpm	0 gpm
LP Compressed Air	160 scfm	0 gpm	0 gpm
Electronic Dry Air	4 scfm	0 gpm	0 gpm
HVAC			
Air Conditioning	35 tons	14 tons	30 tons
Replenishment Air	180 cfm	150 cfm	75 cfm
Purge/Blow-out Air	1,380 cfm	1,200 cfm	1,320 cfm

SOURCE: P. Beurman, Lew, and Webster, 1996, pp. 3–14.

NOTES: gpm = gallons per minutes; WDCM = washdown countermeasures; scfm = standard cubic feet per minute; cfm = cubic feet per minute.

[a] The connected loads listed in the VLS point design column are based on data available at the time of the source report.

ifications that were considered the "most demanding" for this zone. The second column is the specifications for an "SSES point design for VLS" and the final column reflects DDG-51 contracted specifications as understood at the time of the source report. We draw the reader's attention to two highlights from this table. The first is that for the electrical loads the DDG-51 implemented the full SSES specifications. The second is that in cases of potable water, high-pressure air, low-pressure air, and electronic dry air, the lower specification was part of the contract. In practice, this essentially means that although sometimes touted as a modular installation, the DDG-51 VLS was only a preliminary implementation, the only guidelines that were available at the time. The ability to insert a different or upgraded weapon module into the VLS weapon design zone that did not adhere to the design constraints of the VLS could require extensive redesign, making a future upgrade more costly and potentially less likely to occur.

The Aegis Combat System in DDG-51 Design

The second example of modularity initiatives related to the DDG-51 considers the inclusion of the ACS in the DDG-51 design. Although not often directly associated with previous modularity programs, ACS was inherently modular in both hardware and software design. Figure B.2 shows an example from the mid-1980s of a system (rather than architecture) perspective of the ACS. We contend that this representation reflects the manner in which the program office of that time wished the program to be viewed: as a combination of systems that would become an integrated multi-mission weapon system.

How has this modularity initiative been executed by the U.S. Navy in the context of DDG-51? The Navy has had the ability to develop and insert new capabilities during new construction and modernization through three flights and ACS upgrades. Each ACS upgrade, historically called a baseline (BL), is composed of both computing hardware and software improvements. Figure B.3 provides additional insight into the effect of a modular combat system. The figure shows the initial BL inserted into the DDG hulls at the bottom as well as the baseline the hull supports currently. Our assertion is that although the DDG-51 was not a full modular design, the ACS has modular princi-

Figure B.2
Overview of Aegis Ship Combat System from 1986

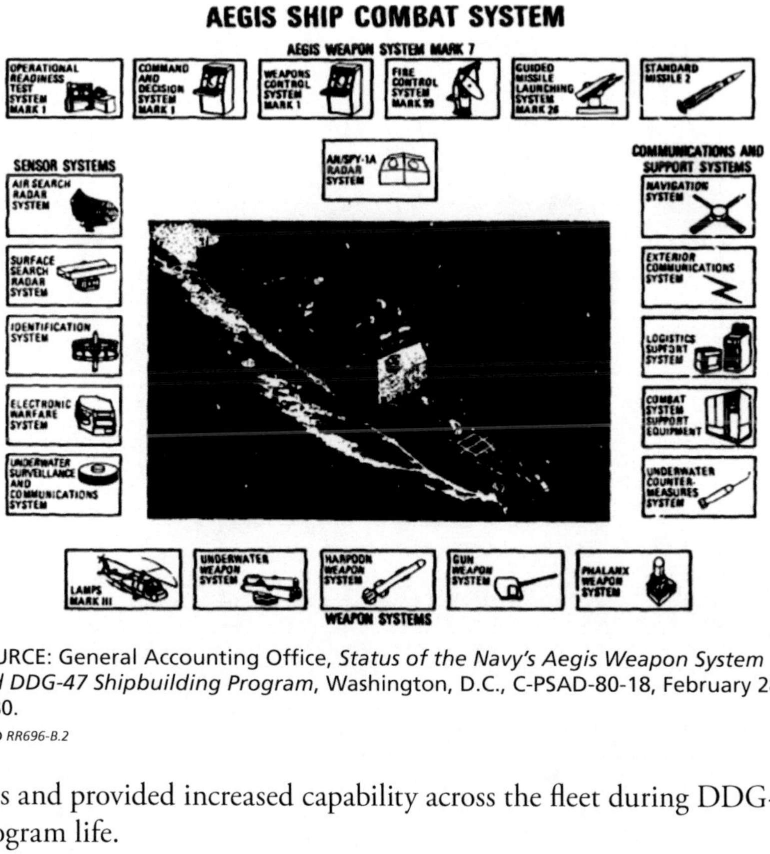

SOURCE: General Accounting Office, *Status of the Navy's Aegis Weapon System and DDG-47 Shipbuilding Program*, Washington, D.C., C-PSAD-80-18, February 28, 1980.

RAND *RR696-B.2*

ples and provided increased capability across the fleet during DDG-51 program life.

Did the Lack of a Total Modular Adaptable Ship Design Keep Important Capabilities from Entering the Fleet?

We next turn our attention to our second issue as part of this assessment: Were there important capabilities unable to be inserted into the design because the ship was not a total MAS design? The best example of a delayed capability is the gun weapon system zone for the DDG-51. The intent of the SSES program of the early 1980s was to include a

Figure B.3
Aegis Weapon System Baseline Capabilities

BL 4.1	BL 4.2	BL 5.1	BL 5.3	BL 6.1	BL 6.3	BL 7.1
AN/SPY-ID	MK 36 DLS MOD 12	OJ-663 consoles	RLGN	AN/SQQ-89(v)10 Kingfisher	Area TBMD SM-2 BLK	SQQ-89(v)15
UYK-43-44	Remote optical	SM-2 BLK III	JTIDS/CP/LINK 16	VLS B/L IV	IVA	Area TBMD
MK 34 GWS	Sight.	SM-2 BLK IV	Model 5	FODMS	CADRT	SPY-ID(V) w/ORTS mods
SPS-67(V)3	Advanced LSD	SPY computer program	Tactical graphic	BFTT (ACTS rehost)	CEC 2.0 (AAW)	Advanced processing
ADS MK II	LINK II upgrade	Upgrades	capability	Raised SPY arrays	JMCIS (RX, MDX)	AN/WLDj-1 RMS
WRN-G GPS	ASW OBT	Training upgrades	UYH-16 MMSD	NSFS	ALIS	AIEWS increment 1
SM-2 BLK II	SSTD F1	CASS upgrades	Combat DF	MK 34	CDLMS	VLS baseline 6 upgrade
MK 116MOD7	VLS	ECPs:	TADIXS B/FRE	Mod 1 GWS	NAVSSI(BLK 4)	OJ-191 replacement
TWS SWG-SA B(KII)	Plasma computer	INTPS	ATWCS	MK 160 Mod 8	C2P rehost	MK 34 Mod 1
DMS	controlled	NTDS serial controller	SLQ-32(V)3	GCS	AN/SQQ-89(V)14	GWS/MK 160
SYQ-7(V)5/54	Entry panel	improvement	OJ-663(TOG)	MK 45 Mod 4	RSCES	Mod 9 GCS
(NAVMACS)	TWS/ASW	CIWS sum and Delta	Computing suite	Gun	BFIT (AWS integration)	Navy link certification
ECPs:	Integrated training	Range mod.	upgrades	Mount ERGM	AOCD DX/DR upgrade	TTWCS (DDG 96+)
LLSI	TSS submode		SPY-ID w/TIP	NFCS	ESSM/FCS	VLS B/L 7 upgrade
SQQ-89 DS	ECPs:		RCS TURNKEY	ORTS upgrade	STAMO/VLS	(DDG 96+)
SGS/AC	CIWS		ECPs:	AWS improvements	B/LS	COTS refresh
	Pneumatic gun drive		MK-162 Mod 0 (SGS	ATWCS LCGR	Navy link certification	Hawklink C to KU band
			computer)	ADS MK 6 CLSD		Conversion (DDG 96+)
			USH-32	DSVL		
			Recorder/reproducer	Armed helo capability		
			upgrade	Dual helo hangar &		
				SH-60 B/R		
				Q-70 (DDG 81+)		
				Link-16 certification		
				Consolidate peripheral		
				optical disk AOCS		
				(EMULATE mode)		
				"A" scope synthetic display		
				Delete TACSAS, Harpoon,		
				VLS loader crane		

DDG 51–53

DDG 54–58 — Current

DDG 59–67

DDG 51–53
DDG 54–58
DDG 59–67
DDG 68–78

DDG 79–84

DDG 79–84
DDG 85–90

DDG 91–102

Initial

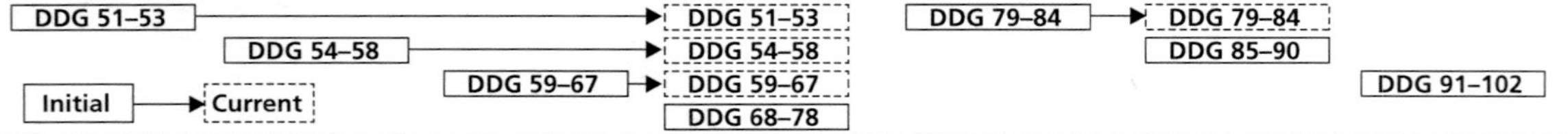

SOURCES: Based on previous RAND research on the Navy's Aegis program. Historical data provided by Randy Fortune, former U.S. Navy DDG Program Officer.

NOTE: Later baselines include capabilities of all previous baselines. Baseline plan dated April 2000. Current DDG 103–112 initial baseline designated BL 7.1R. Details not part of reference document.

RAND RR696-B.3

weapon zone adhering to ship system engineering standards that would have initially contained the 5"/54 caliber gun. Similar to the VLS weapon zone, initial guidance and design requirements were established and the preliminary engineering standards developed. A major difference between the gun system and the VLS was that a 5"/54 caliber gun *module* (designed for the weapon zone module designated AA) was not fully developed. Beurman, Lew, and Webster (1996) indicates that a fully engineered and validated module *did not exist*. Under Chief of Naval Operations directives in November 1983 that targeted the reduction of cost and weight, the decision was made to install a conventional gun system on DDG-51 instead of one with SSES standards.

The idea of including a modular gun weapon zone did not lie dormant. The issue was revisited in 1996 with the development of an engineering change proposal (ECP) by Gibbs & Cox, Inc. for the DDG-79 hull, described in the following way:

> A Modular 5"/54 Gun System (MGS), consisting of a Blohm and Voss 5"/54 Gun Module and a standard 5"/54 caliber United States Navy gun, will replace the existing 5"/54 caliber gun.[4]

This ECP was a very extensive engineering change, involving 175 pages of specific contract-level specification changes and ship drawings. Eighteen individual specifications were altered and 24 ship drawings were affected. Many more construction (fabrication) documents would be affected. To provide a sense of the extent of the ECP, the affected drawing titles (as referenced in Gibbs & Cox) were as follows:

- General Arrangements—Inboard Profile and Sections
- General Arrangements—Main Deck and Below
- General Arrangements—01 Level and Above
- General Arrangements—Topside Configurations
- Gun, Torpedo and Missile Weapon System Capabilities Block Diagram

---

[4] Gibbs & Cox, Inc., *Draft Engineering Change Proposal for MGS Installation on DDG 51 Flight IIA Class Ships*, January 24, 1997.

- Combat Sys Equipment Room No. 1 and Sonar Control Room—Arrangement of Equipment
- Voice Interior Communications Systems
- Alarm and Indicating Systems
- 5"/54 Caliber Loader Drum Room—Arrangement of Equipment
- CIC and Sonar Control Room Lighting
- Heating, Ventilation, and Air Conditioning Design Criteria Manual for the DDG-51 Class
- DDG-51 Class, Heating, Ventilation, Air Conditioning Diagram
- High Pressure Air System Diagram
- Nitrogen System Diagram
- Air Conditioning Chilled Water Circulation Systems and Condensate Drain Diagram
- Sprinkler System Control Valves Thermo Pneumatic and Hydraulic Piping Diagrams
- Firemain System Diagram
- Magazine Sprinkling System Diagram
- Main and Secondary Drainage System Diagram
- Aqueous Film Forming Foam Sys
- Washdown Countermeasure System Diagram
- Electronic Cooling Water Sys
- Ship Service and Dry Air System Diagram
- 5"/54 Caliber Gun Module Foundation and Supporting Structure.

The primary effect of a conventional compared with a modular gun system (both at the initial and Flight IIA design, as discussed above) with respect to fleet capability may have been to delay—until hull 81—insertion of the 5"/62 gun system into the design. During the 1980s and 1990s, the Navy expected and desired enhanced capabilities from the Extended Range Guided Munition (ERGM). Although production of that weapon ultimately did not occur (the program was cancelled in 2008), the 5"/62 gun is still a lighter-weight, more-easily maintained gun system with a longer barrel, improving gun effectiveness during naval surface fire support.[5]

Did Modularity Lead to Increasing Acquisition Costs?

One of the promoted benefits (although perhaps not the primary benefit) of modularity from our literature review was the potential to reduce new construction costs as well as accelerate the implementation of advanced technology (which provides the benefit of more effective ships). This leads to the question of how the DDG-51 program performed (with respect to acquisition costs) as capability improvements were introduced into the fleet. Did those increasing capabilities lead to increasing acquisition costs? Figure B.4 provides the DDG-51 acquisition profile (based on data from its Selected Acquisition Report) for the hulls procured through 2005. As the figure indicates, per-unit costs have been within $650–950 million (for base year 1987 dollars) during this time frame, and, as indicated earlier, the Aegis baseline program has provided increasing capability.

Concluding Remarks

In this appendix, we examined the DDG-51 program as a case study for assessing modularity in surface ship design. This case study was performed as a macro-assessment of several key issues uncovered during the course of this research. We focused our assessment on two issues:

- What were the previous flexibility and modularity initiatives and plans? How were the plans executed?
- Did the lack of a total MAS design keep important capabilities from entering the fleet?

---

[5]   U.S. Navy, "U.S. Navy Fact Sheet: MK 45—5-inch-gun 54/62 Caliber Guns," *Navy.mil*, October 19, 2012

Figure B.4
DDG-51 per-Unit Cost and Quantity

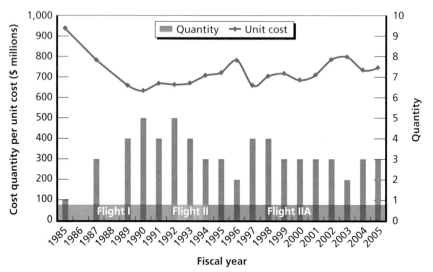

SOURCE: Department of Defense, 2011b.
RAND *RR696-B.4*

Table B.2 provides a summary of this assessment. We characterize our case study as a "mixed bag." For example, modularity was clearly considered during the original DDG-51 program as a result of the SSES program with the VLS system. But it is also true that the weapon zone containing the VLS was built to only preliminary SSES specifications. Has this affected the DDG-51 in any way? The answer is uncertain; VLS has stood the test of time as an effective weapon system, and the Navy has not yet needed to replace that weapon zone. Additionally, the ACS has modular principles and has provided increased capability (warfighting effectiveness) over the program life.

What else does the DDG-51 experience tell us with respect to flexibility and modularity? Our research suggests the following items for consideration:

- Modularity principles have helped, but the DDG-51 design could have gone further had additional SSES standards been available.

Table B.2
Case Study of the DDG-51 Program Is a "Mixed Bag"

Issue	Response
What were the previous flexibility and modularity initiatives and plans? How were plans executed?	Modularity considered during DDG-51 design in SSES Only forward and aft VLS cells had SSES standards developed Though not a full modular design, the ACS has modular principles and provided increased capability over program life
Did a lack of a total MAS design keep important capability from entering the fleet?	Consideration given to a modular gun installation—not done as it provided no additional combat capability at a significant increase in cost Program commitment gave opportunity to insert capability by means of changes to design (e.g., 5"/62 gun in Flight IIA for DDG-81)

- When considering design changes, increased weight or volume is viewed as limiting factor because of naval architectural limits due to weight-based cost modeling.
- Acquisition cost effects from modularity are uncertain.
- Some studies suggest increased acquisition costs from higher material costs.
- Others promote the notion that acquisition costs will be lower due to reduction in labor hours/time in shipyard.
- There is agreement on savings in total ownership costs, largely from recoupment during modernizations over time.
- New technologies/capabilities and threat evolution may require new design regardless of the level of modularity.

Flexible Infrastructure

Huntington Ingalls Industries–Newport News Shipbuilding and the U.S. Navy Aircraft Carrier Program have embarked on a Flexible Infrastructure Program. It is aimed at providing the capability to quickly upgrade/modernize or re-fit a system/compartment while eliminating the need for hot work. The project stems from the continuing aircraft carrier mission system development during the longer aircraft carrier construction period and the challenge to install, integrate, and test mission systems late in the construction period. The system, with its pre-installed infrastructure, seeks to eliminate the conflicts stemming from change orders and mission development during construction, and the associated cost and schedule impacts. The system appears to borrow concepts from aircraft designs. It is composed of seven key elements:

- A deck track system, with three heights (6 inches, 9 inches, 12 inches) for various compartment applications. The deck track includes standard deck tiles mounted in the deck track (Figure C.1)
- A bulkhead track system (Figure C.2)
- An overhead track system (Figure C.3)
- Portable bulkheads and stanchions (Figure C.4)
- An under-deck ventilation system composed of a 12-inch deck track plenum system and a 6-inch deck track hybrid system. Separate supply rooms and exhaust plenums are required for this system
- Flexible lighting
- Flexible power.

Figure C.1
Flexible Infrastructure Six-Inch and Twelve-Inch Track System

SOURCE: Huntington Ingalls Industries–Newport News Shipbuilding.
RAND *RR696-C.1*

Figure C.2
Flexible Infrastructure Bulkhead Track System

SOURCE: Huntington Ingalls Industries–Newport News Shipbuilding.
RAND *RR696-C.2*

The deck track system is fabricated from extruded aluminum and installed on studs welded to the deck. Deck tracks are mounted athwartships on 12-inch centers, with leveling nuts placed on the studs to adjust for deck unevenness and to provide a level track surface. Attachment points are located every inch along the deck track for bolting equipment foundations at various positions. The attachment points

Figure C.3
Flexible Infrastructure Overhead Track System

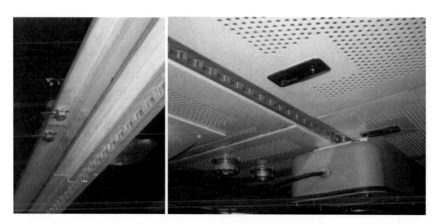

SOURCE: Huntington Ingalls Industries–Newport News Shipbuilding.
RAND *RR696-C.3*

are referred to as "track profile." The bulkhead track system is located on two-foot centers. The overhead track system is similar to the deck track, except it is also located on two-foot centers. It has similar attachment points. The compartment design elements are individually tailored to meet specific compartment requirements.

Newport News Shipbuilding has installed its flexible infrastructure in several combat system spaces on the USS *Gerald R. Ford* (CVN-78) to support mission systems. The flexible infrastructure system is shock qualified for the specific location on the *Ford*.

Flexible infrastructure is potentially a significant enabler for surface combatant modernizations, and particularly for capability insertions outside of scheduled modernization availabilities. However, while the engineering has been completed for use on *Ford*, further engineering is required for widespread use on destroyers. Some of the technical issues that must be addressed include the following:

- *Corrosion/bi-metallic interaction.* Flexible infrastructure currently uses an aluminum alloy for the deck and overhead track. Use of aluminum on steel decks will require testing to ensure compatibility and resistance to fatigue.

Figure C.4
Flexible Infrastructure Stanchions and Bulkheads

SOURCE: Huntington Ingalls Industries–Newport News Shipbuilding.
RAND *RR696-C.4*

- *Bonding and grounding.* Bonding and grounding has been a challenging shipboard issue. With the new mounting, it must be addressed for flexible infrastructure—in concert with the above corrosion issue.
- *Shock and vibration.* Use of flexible infrastructure on destroyer-class ships will subject the system to a different shock and vibration regime depending on the system's location on the ship. Thorough engineering and testing needs to be competed for flexible infrastructure use on destroyers.
- *Through-life maintenance.* Flexible infrastructure is a departure from previous installation approaches. Standard drawings, technical manuals, and logistics support are needed.

- *Ship Specifications.* In addition to the above, flexible infrastructure must comply with appropriate ship specifications.

While flexible infrastructure is potentially a significant enabler for destroyer-class modernizations, by itself it is not sufficient. Access to the affected space through welded access panels should be considered part of the design. Further, ship system services; piping; electrical cable runs; and heating, ventilation, and air conditioning (HVAC) trunks should be located to provide clear access to the affected compartments.

USS *John Paul Jones* (DDG-53) Mid-Life Modernization

Table D.1
Description of DDG-53 Modernization Work Packages

Requirement	System	Execution Impact
74012	VLS Magazine Ventilation	*Justification:* With this change, ships will have the required additional ventilation for SM-3 missile loadout. *Ship Impact:* This change upgrades ventilation capacities in the forward and aft VLS magazines for ballistic missile defense (BMD)–capable DDGs. SM-3 missiles require additional cooling in VLS; current HVAC conditions in the magazines are inadequate for maintaining requisite launch temperatures for expected solar heat load and environmental conditions. All four fan rooms servicing the forward and aft launchers will be completely changed out. Electrical upgrades will be accomplished as needed for fans and heaters.
75928/82635	Circuit 3TV	*Justification:* Unsupportable and inoperable CKT 3TV system will not provide the visibility required in today's environment to support ships' force safety and overall situational awareness. *Ship Impact:* DDG Modernization Advanced Capability Build 12 (ACB12) TOPSIDE SURVEILLANCE/CAMERA COTS UPGRADE: CKT 3TV System was implemented to provide surveillance for VLS/helicopter operations. Current CKT 3TV systems components are obsolete and unsupportable, which leads to an inoperable system that does not provide the situational awareness and adequate coverage for the VLS/helicopter operations.

Table D.1—Continued

Requirement	System	Execution Impact
76829	SPA 25-H	*Justification:* Without this change, ships will have increased maintenance and failures due to obsolete equipment remaining on board. *Ship Impact:* Addresses obsolescence with the replacement of the legacy AN/SPA-25G Indicator Group with the new AN/SPA-25H Indicator Group. The AN/SPA-25H Indicator Group is an advanced navigation and tactical situation solid-state radar indicator for both Combat Information Center (CIC) and ship bridge environments. The system is a variant of the U.S. Navy AN/UYQ-70 program that provides computer-based radar display consoles developed for use on Navy shipboard applications.
76869	Announcing System 1MC	*Justification:* The current General Announcing System consists of historic difficulties that need to be overcome. A few include single points of failure in the area of amplifiers and the absence of survivability. Additionally, this system does not allow for any expansion to accommodate a modernized ship. *Ship Impact:* This change will replace the General Announcing System equipment, which consists of an amplifier assembly AM-2316, amplifier oscillator group AN/SIA114, and amplifier control groups AN/SIA-119A and AN/SIA-120. The equipment will be replaced with a redundant dual center system comprising two Dynalec Integrated Announcing System (DIAS) cabinets and two Dynalec Cross Connect Field (DCCF) units. All loudspeakers will be reused; however, the existing 1MC terminal boxes will be replaced with Smart Terminal Junction Boxes.
77052	Internal Voice Communication System	*Justification:* The current AN/STC-2(V) is not economically expandable and has reached its end of life, requiring extensive maintenance to keep it operational. It is currently out of production and logistics support is quickly becoming unavailable. *Ship Impact:* This change will replace AN/STC-2(V) with an AN/STC-3(V)2 (SHINCOM IVCS) System. Replacement of AN/STC-2 System with SHINCOM IVCS will introduce a common communications infrastructure integrating interior administrative and tactical systems with exterior communications.

Table D.1—Continued

Requirement	System	Execution Impact
77615	Open Architecture Computing	*Justification:* Without this change, DDGs 51–78 will be unable to perform Anti-Air Warfare (AAW) missions in ACB12/Technology Insertion 12 (TI12) (ACB12/TI12). *Ship Impact:* This change will deliver and install the equipment related to the core Aegis Weapon System (AWS) CR3 OA Computing Plant Upgrade on DDGs 51–78 as part of ACB12/TI12. The scope of this change covers equipment primarily related to the AWS CR3 computing plant. This change removes/replaces existing AWS computing hardware and Secure Voice System and delivers and installs the following: Common Processor System (CPS) cabinets, Aegis Conversion Equipment Group (ACEG) cabinets, Navy Tactical Data System In/Out (NTDS I/O) cabinets, Aegis LAN Interconnect System (ALIS) cabinets, Thin Client Displays, COTS Printers, LAN Radar Data Distribution System (LRADDs), and Secure Voice System.
78391	SQQ-89A(V)15 w/ MFTA	*Justification:* Without this change, ships will not receive the latest Sonar System containing OSA and numerous other improvements. *Ship Impact:* The AN/SQQ-89 ASW Combat System provides an integrated Undersea Warfare capability for the ACS. This change upgrades the ASW system to the AN/SQQ-89 A(V)15 w/ EC-211, including a Multi-Function Towed Array (MFTA) with integrated Acoustic Intercept (ACI) and removes the AN/SQQ-89(V)4 ASW (DDG-51) and AN/SQQ-89(V)6 Block I ASW (DDGs 52–78) systems.
78511	CIC Display Upgrades	*Justification:* This change provides necessary elements for antiterrorism/Force Protection; replaces AN/UYQ-21 family equipment that has become unavailable, obsolete, or unreliable; reduces total ownership cost through lower material costs and maintenance hours; and is part of the Warfighting Improvement Plan. This upgrade is integral to ACB12/TI12. *Ship Impact:* This change will deliver and install new equipment, and will rearrange existing equipment, to implement the redesign of CIC on DDGs to address the utilization of the new Common Display System (CDS) consoles and video technology as part of ACB12/TI12. The rearrangement of CIC is required due to an increase in console size versus current AN/UYQ-21/70 console variants. This CIC rearrangement/redesign effort also provides an opportunity to recapture space currently occupied by existing rear projector displays, to address future capabilities and warfighting improvements and to optimize watch-stander functionality.

Table D.1—Continued

Requirement	System	Execution Impact
78512	Multi-Mission Signal Processor	*Justification:* Without the MMSP installed, DDGs 51–78 will be unable to perform anti-air warfare and BMD missions in ACB12/TI12. *Ship Impact:* This change will replace the legacy SPY-1D Signal Processor Group, OL-356/SPY-1D with the new MMSP. The MMSP is a critical enabler for ACB12/TI12 System, as it provides SPY-1D(V) anti-air warfare and BMD 5.0 capabilities, utilizing and leveraging existing Aegis BMD Signal Processor (BSP) designs. MMSP provides COTS Signal Processing architecture, reduced footprint and acquisition cost.
78513	Kill Assessment System (KAS)	*Justification:* Without KAS Warhead Data Receiver Cabinet (WDRC) installed, DDGs 51–78 will be unable to perform BMD missions in ACB12/TI12. *Ship Impact:* The below-deck portion of KAS consists of two WDRCs housed in Mission Critical Enclosures. These interface with the Fire Control System Director and Weapon Control System Processing housed in the TI12 computing infrastructure. For DDG 51–78 ships that have BMD 4.0.1 capability prior to receiving ACB12/TI12, this change will modify existing MK 78 MOD 0 WDRC cabinets to the MK 78 MOD 1 version by installing the single latch door. In addition, BMD 4.0.1 ships will have the Mission Planner Consoles used as KAS displays removed, since they are not need in ACB12/TI12.
78819	Gun Weapon System Upgrade	*Justification:* Gun Weapon System (GWS) inability to work in the CDS/CPS/OA environment and incompatibility with interfaces and requirements specified for GWS in the ACS Specifications for ACB12/TI12. *Ship Impact:* The intent of this change is to upgrade the Mk 34 Mod 0 GWS to Mk 34 Mod 7 by replacing the existing Mk 160 Mod 3 (DDGs 51–73)/Mod 6 (DDGs 74–78) Gun Computer System with the Mk 160 Mod 15 Gun Computer System and replacing the Mk 46 Mod 0 Optical Sight System with the Mk 20 Mod 0 Electro Optical Sight System.

Table D.1—Continued

Requirement	System	Execution Impact
79256	CEC ANUSG-2B	*Justification:* The AN/USG-2B Cooperative Engagement Capability (CEC) system is installed on applicable ships to enhance and improve the overall functional capabilities of the Expeditionary Strike Group by allowing for engagement coordination between cooperating units. CEC is based on distributed generation of a database of merged sensors, target identification, and decision data for the purpose of common force-wide information for pursuing coordinated and cooperative use of weapon systems. If this change is not accomplished, there will be no capability for the Battle Group (BG) to send/receive real-time information and data exchange, and allow for engagement coordination between cooperating units of the BG. *Ship Impact:* The CEC installation consists of the CEC Processing Group, AN/USG-2B, and the Antenna Assembly, AS-4558/USG-2. The Processing Group equipment is mounted in the Communication Processing Set, MT-7292/USG-2B (LC 28-01-51). The AS-4558/USG-2 Antenna Assembly consists of four Polyalphaolefin (PAO) liquid-cooled antenna arrays connected in series. These antennas are responsible for transmitting and receiving CEC data.
79584	VLS Upgrade	*Justification:* If this change is not accomplished, VLS will not be able to support the next-generation Evolved Sea Sparrow Missile and Aegis BMD 5.0 programs, which includes the capability to fully support the SM-3 Block IA/IB missile variants. *Ship Impact:* The purpose of this change is to replace both VLS AN/UYK-44–based Launch Control Units (LCUs) with two Q-70-based Mk 235 Mod 6 LCUs to provide both BL III and VII functionality (two per ship). The new Q70s will incorporate VLS Global Positioning System (GPS) Integrator (VGI) capability that was resident on previous VGI racks. The COTS-based Versabus Module European (VME) enclosure will house the LCU processor and the functional modules of the Advance LCU Peripheral (ALP). The VLS Data Terminal Group and Signal Data Recorder/Reproducer Set equipment will be removed as part of this change. One standard module in each launcher is upgraded to BL VII (F8 and A8).

Table D.1—Continued

Requirement	System	Execution Impact
70403	Digital Fuel	*Justification:* If the alteration is not accomplished, ships will continue to refuel using the antiquated panels they currently have and therefore 1. There will be a continued high risk of fuel spill during refueling 2. The refueling operations will continue to be manpower-intensive 3. Maintenance and calibration costs of the existing system will continue to rise 4. Ship fuel data will not be automatically available to the Battlegroup 5. Liquid load reports will continue to be taken manually. *Ship Impact:* This change implements a fuel fill and transfer control system that will provide improved control during refueling and fuel transfers at high fill rates to a maximum number of tank groups with a reduced risk of overpressurization of tanks and/or fuel spills. This will allow for shorter alongside times during underway replenishment and also increases the operation, maintainability, and reliability of the fuel system. Networked system will allow automatic reporting of fuel capacities and burn rates to determine mission capabilities. This change also creates an auto fill feature of the Gen 3 Head Tanks.
71604	Machinery/ Damage Control System Upgrades	*Justification:* This change provides the necessary hardware and software modifications to reduce manning by allowing the engineering plant to be operated by a single enlisted watchstander and the Engineering Officer of the Watch. *Ship Impact:* Single Central Control Station watchstander is one of the core DDG modernization changes being implemented on forward fit on DDGs 111 and 112. A "Universal Control Console" (UCC) approach is employed, which allows propulsion and electric plant operation from a single console by one person. The two primary UCCs will be in the Central Control Station in the current location of the PACC and EPCC consoles. UCCs will also exist in the two engine rooms, taking the place of the Shaft Control Units (SCUs).

Table D.1—Continued

Requirement	System	Execution Impact
71605	Advanced Galley	***Justification:*** This alteration updates the food service space arrangements and equipment and leverages improved food service technologies currently aboard other Navy ships. Rearrangement of the food service spaces and replacement of obsolete equipment will facilitate S-2 Division manning reduction through faster cooking times, multi-functionality, increased capacity, reduced clean-up times, improved process flow within the galley, and use of advanced food packaging/production methods. ***Ship Impact:*** If this alteration is not accomplished, DDGs food service system will continue to be labor-intensive, will not adequately support preparation of the new advanced foods, hinder quality of life/personnel retention, and will not reap the life-cycle cost benefits of reduced manning of the S-2 Division.
71726	Full Integrated Bridge Navigation System Upgrade	***Justification:*** Reduces manning and allows the ship to be operated with the Officer of the Deck (OOD), Junior Officer of the Deck (JOOD), and one enlisted watchstander. ***Ship Impact:*** The single bridge watchstander change is one of the core DDG modernization upgrades being executed during forward-fit on DDGs 111 and 112 and installs the Integrated Bridge and Navigation System (IBNS) with the necessary hardware and software changes to allow operation at reduced manning levels, with a three-person bridge team. The primary operating station is moved from the existing Ship Control Console (SCC) to a "Helm Forward" station positioned directly below the forward windows of the Pilot House. The SCC is retained as a backup operating station. The Helm Forward design includes four operating stations under the windows including ARPA, Steering/Propulsion Control (SPCS), VMS, and the Pilot House Machinery plant/Repair Station Console (MCS/RSC).

Table D.1—Continued

Requirement	System	Execution Impact
73088	Gigabit Ethernet Data Multiplex System	*Justification:* The Data Multiplex System (DMS) is the mission critical, general purpose, control data, first-generation shipboard network for DDG-51 Class Destroyers. This network handles inputs and/or outputs from the Machinery Control System (MCS), Damage Control System (DCS), Steering Control System (SCS), Combat System, Navigation Displays and Interior Communications (IC) Alarms and Indications. The copper RF-based DMS network installed on DDGs 51–78 is becoming increasingly obsolete. The installation of a Gigabit Ethernet Data Multiplex System (GEDMS) is the only feasible solution for replacing DMS in support of DDG modernization. *Ship Impact:* 1. Insufficient bandwidth for the increased data and video throughput required to achieve significant reductions in watch-standing requirements, as essential step to reducing DDG crew size. 2. Ships will not benefit from cost reductions of reduced manning. 3. RF components are becoming increasingly obsolete. Costs for parts are increasing while availabilities of parts are decreasing. 4. Time, cost, and level of effort for maintenance, troubleshooting, and repair will continue to increase.
73622	Hydrogen Sulfide Detectors	*Justification:* If hydrogen sulfide detectors are not installed, more injuries, or possibly deaths, could result on DDG-51 class ships. *Ship Impact:* This change installs two hydrogen sulfide gas detection systems, one in each VCHT Room, and one hydrogen sulfide calibration kit. The systems will include local and remote visual and audible alarms. The detectors will provide a means to notify personnel of the presence of dangerous hydrogen sulfide gases within the VCHT Rooms; civilian and military personnel injuries have resulted from not having the detectors, and deaths associated with hydrogen sulfide in shipboard sewage systems have occurred.
76034	Emergency JP-5 Fill Capacity	*Justification:* During a review of the USS *Cole* lessons learned it was determined that having the JP-5 service system providing an emergency source of fuel to SSGTG No. 3 head tanks proved beneficial. This DDG MLE alteration adds a backup fill capability to the SSGTG No. 3 head tanks. *Ship Impact:* During another USS *Cole*–like event, ships will not be able to fill their SSGTG No. 3 head tanks from the JP-5 service system (JP00).

Table D.1—Continued

Requirement	System	Execution Impact
76186	UNREP Midship Padeye Relocation	*Justification:* UNREP procedures will continue to be manpower-intensive and SPY-1D operations will continue to be curtailed during UNREP operations. *Ship Impact:* This change modifies the midships sliding padeye installation. On DDGs 51–58, the padeye is stowed at a 35-degree angle on a fixed stowage kingpost. During UNREP operations, the padeye is raised to vertical and support by backstays, which are attached to the forward and afterdeck houses. This ship change relocates the sliding padeyes six feet inboard and sets them at a permanent ten-degree angle with fixed backstays. The ship foundation plan is revised to make the ten-degree tilt. The SPE backstays to be revised. Power must be revised to suit. DDGs 59 and later had this change accomplished during new construction via ECP 51-0950R1C1. This change will allow the SPY-1D radar to be operated even when UNREP operations are in progress. The midships sliding padeye UNREP setup operations will be eliminated.
76253	MRG/CPP/L-O REPL w/ Filter SEP	*Justification:* The current purifiers (Alfa Laval) have experienced frequent failure with motors, belts, shafts, seals, valves, hose spindles, and switches. If ship installation is not installed, the system will be status quo. *Ship Impact:* The current centrifugal purifiers (Alfa-Laval) have experienced frequent failures with motor, belts, shafts, seals, valves, hose spindles, and switches. The new filter/separator (F/S) units have far fewer parts; no freshwater tank, controller, sludge discharge, or resilient mounts. The F/S requires less connecting piping, hangars, hoses, maintenance equipment, and tools. They are quieter and weigh less. These lube-oil F/S units are similar to that presently in service on *Seawolf*-class submarines, and planned for installation aboard *Virginia*-class submarines and CVNX carriers.

Table D.1—Continued

Requirement	System	Execution Impact
76648	Reefer Plant Overboard Pressure Relieving	*Justification:* This alteration addresses a potential safety issue with the ship stores refrigeration plant for DDGs 51–78, where an over temperature situation could result in a material failure of the plant's condensers. This would cause significant release of refrigerant into a confined space and possible personnel injury. *Ship Impact:* Currently there is no pressure-relieving system for the condensers. If the temperature in the vicinity of the secured unit were to rise above 146 degrees Fahrenheit, the 250 PSIG pressure rating of the receiver and condenser would be approached. Absence of an overboard pressure relieving system for each of the ship stores refrigeration plant condensers could lead to explosion, personnel injury, and significant release of refrigerant into a confined space.
76974	Adds Moriah Wind System	*Justification:* If this alteration is not accomplished, DDGs will continue to use the legacy Type F Wind Measuring and Indicating System (WMIS), which utilizes outdated 60-year-old syncho-mechanical technology. WMIS is labor-intensive, with excessive calibration and maintenance requirements and high costs for obsolete parts. Most importantly, WMIS is incapable of interfacing with new shipboard digital systems and will not support the goal of the DDG Modernization Program and Sea Power 21. *Ship Impact:* This alteration adds the Moriah Wind System (MWS) to the DDG-51 ship class, pursuant to the DDG Modernization Program. MWS is an ACAT IVM program managed by NAVAIR PMA-251. MWS is in the Modernization Plan and is an approved legacy alteration for CVNs, LHAs/LHDs, and CGs, and is also being installed on new-construction DDG102AF. MWS provides digital wind speed and direction information, including crosswind and headwind, that supports decisionmaking for air operations and combat. In addition, MWS displays LREs and VSTOL Bulletin Data.

Table D.1—Continued

Requirement	System	Execution Impact
77259	Replace Legacy EM LOG W/AN/ WSN-9(V)1	***Justification:*** Failure to fund this effort will result in Flight I and Flight IIA DDGs continuing to use a speed log that is currently experiencing a lack of availability of replacement components, exacerbating an already bleak supportability issue, requires frequent calibration, and is difficult to maintain. ***Ship Impact:*** This SCD upgrades DDG-51 Flight I and Flight IIA Own Ship's Speed Log from the MK 4 series or MK 6 series of EM Logs to AN/WSN-9(V)1 Digital Hybrid Speed Log (DHYSL), the Program of Record Own Ships Speed (OSS) Log per PEO IWS 6. The Electromagnetic Log (EM Log) is a component of the conventional navigation system used aboard naval surface ships and submarines. EM Logs operate in conjunction with a hull-mounted sensor to measure ship's speed relative to the water and distance traveled from a given starting point.
77269	Replace Redundant Tank Level Indicators	***Justification:*** Based on lessons learned from USS *Cole*, visual fuel indicators are required for the three SSGTG fuel oil head tanks in the event of loss to the Fuel Control Console in the Test Lab. In the case of the *Cole*, the crew was forced to use JP5 to fill the SSGTG No. 3 head tanks but had no level indication to judge the amount of fuel in the tank or the amount being used. As a result, a tank was run dry, starving the SSGTG. The *Cole* was without electricity for 21 hours. ***Ship Impact:*** This change installs GEMS Suresite Tank Level Indicators (TLI) on the SSGTG Fuel Head Tanks to provide redundant capability to visually verify the tank levels and rate of fuel usage in the event of loss of the Fuel Control Console in the Test Lab. This change is a lesson learned from the USS *Cole* as tanks are likely to run dry, starving the SSGTG and resulting in a loss of electricity.

Table D.1—Continued

Requirement	System	Execution Impact
77419	Textile Ducting	*Justification:* Textile Duct is a flexible, porous Nomex material used for supply distribution ductwork. This ductwork supplies air to compartments by inflating a "sock" type section of duct and the air is distributed by passing through the porous material. Use of this type of ducting can result in improved distribution by eliminating traditional supply diffusers that can create objectionable drafts. Textile duct can also reduce airborne noise levels by eliminating flow noise that originates from air passing through the traditional diffuser. *Ship Impact:* PSA conducted installation of Package 512-11-001, on hull 99. The modernization design will mimic the PSA package. The following spaces will have textile ducting installed: pilot house; wardroom, messroom, and lounge; commanding officer cabin; CPO messroom and lounge; crew messroom; central control station; and DC central.
77427	Digital Indicators	*Justification:* To provide situational awareness for watchstanders and removal of obsolete equipment. *Ship Impact:* DDG modernization will not accomplish the goals of improved situational awareness for watchstanders and removal of obsolete equipment.
77829	Install Radar & TDR Tank Level Indicators	*Justification:* Failure to perform will propagate expensive care and maintenance of the existing Gems tank level indicators (TLIs). The average cost to repair a Float is $36,000 per tank and each float is repaired every three years. The average cost to install a Radar TLI is $28,000 per tank. This utilizes money spent fixing floats to provide dramatic future savings. The cost to maintain the Radar has been less than 1 percent of the cost to maintain a Float. *Ship Impact:* This change replaces all float-type TLIs with a Radar or TDR TLIs. The cost to install a Radar or TDR is less than the cost to fix a float. This will provide immediate savings and the future savings are dramatic. The greatest cost to install this alteration is the cost to clean and gas-free the seawater compensated tanks.

Bibliography

Abbott, J., *Modularity Historical Study*, Carderock, Md.: Naval Surface Warfare Center Carderock Division, 2009.

Abbott, J. W., A. Levine, and J. Vasilakos, "Modular/Open Systems to Support Ship Acquisition Strategies," *ASNE Day 2008 Proceedings*, Alexandria, Va.: American Society of Naval Engineers, 2008.

Ackerman, Spencer, "Video: Navy's Mach 8 Rail Gun Obliterates Record," *Wired.com*, December 10, 2010. As of June 24, 2014:
http://www.wired.com/2010/12/video-navys-mach-8-railgun-obliterates-record/

Alkire, Brien, James G. Kallimani, Peter A. Wilson, and Louis R. Moore, *Applications for Navy Unmanned Aircraft Systems*, Santa Monica, Calif.: RAND Corporation, MG-957-NAVY, 2010. As of June 24, 2014:
http://www.rand.org/pubs/monographs/MG957.html

Anderson, Eric J., *Total Ship Integration of a Free Electron Laser (FEL)*, Monterey, Calif.: Naval Postgraduate School, 1996.

Azani, C. H., and K. Flowers, "Integration, Business and Engineering Strategy Through Modular Open Systems Approach," *Defense AT&L Magazine*, January–February 2005.

Bachkosky, J., D. Katz, R. Rumpf, and W. Weldon, *Naval Electromagnetic (EM) Gun Technology Assessment*, Washington, D.C.: Naval Research Advisory Committee, NRAC 04-01, 2004.

Brodie, Bernard, *Sea Power in the Machine Age*, Princeton, N.J.: Princeton University Press, 1941.

Beurman, P., K. Lew, and J. S. Webster, *Ship System Engineering Standards Implemented on DDG 51 Lead Ship Retrospective*, Arlington, Va.: Gibbs & Cox, Inc., August 1996.

Button, Robert W., John Kamp, Thomas B. Curtin, and James Dryden, *A Survey of Missions for Unmanned Undersea Vehicles*, Santa Monica, Calif.: RAND Corporation, MG-808-NAVY, 2009. As of June 24, 2014:
http://www.rand.org/pubs/monographs/MG808.html

Cecere, I. M., J. Abbot, M. L. Bosworth, and T. J. Valsi, "Commonality-Based Naval Ship Design, Production and Support," Naval Sea Systems Command Dahlgren, November 1993.

"CG-21 Guided Missile Cruiser," *GlobalSecurity.org*, undated. As of April 2012: http://www.globalsecurity.org/military/systems/ship/cg-21.htm

Cheung, P., A. Levine, R. Marcantonio, and J. Vasilakos, "Standard Process for the Design of Modular Spaces," *ETS 2010*, Paper No. SNAME-047-2012, June 28, 2012.

Deaton, William A., and James L. Conklin, *Developing Reconfigurable Command Spaces for the Ford-Class Aircraft Carriers*, June 2010. As of February 20, 2014: https://www.navalengineers.org/SiteCollectionDocuments/2010%20 Proceedings%20Documents/ETS%202010%20Proceedings/Deaton.pdf

Defense Science Board Task Force, *Engineering in the Manufacturing Process*, Washington, D.C.: Office of the Under Secretary of Defense for Acquisition, 1993.

Department of Defense, *Unmanned Systems Roadmap 2007–2032*, Washington, D.C., 2007. As of June 24, 2014: http://www.globalsecurity.org/intell/library/reports/2007/dod-unmanned-systems-roadmap_2007-2032.pdf

Department of Defense, *Unmanned Systems Integrated Roadmap FY2011–2036*, Washington, D.C., 11-S-3613, 2011a. As of June 24, 2014: http://www.acq.osd.mil/sts/docs/Unmanned%20Systems%20Integrated%20 Roadmap%20FY2011-2036.pdf

Department of Defense, *Selected Acquisition Report (SAR): DDG-51*, Washington, D.C., December 31, 2011b.

Department of the Navy, *DDG51 (VP) Design Impact Study*, prepared by John J. McMullen Associates, Arlington, Va., Contract No. N00024-82-C-5344, TI-0001-007, TI-0002-0011, January 15, 1983.

Deputy Chief of Naval Operations (Integration of Capabilities and Resources) (N8), *Report to Congress on the Annual Long-Range Plan for Construction of Naval Vessels for FY2014*, Washington, D.C.: Office of the Chief of Naval Operations, May 2013.

DeVries, Richard, Andrew Levine, and William H. Mish Jr., "Enabling Affordable Ships through Physical Modular Open Systems," paper presented at *Engineering the Total Ship (ETS) Symposium* 2008, American Society of Naval Engineers, 2008. As of February 22, 2014: https://admin.navalengineers.org/SiteCollectionDocuments/2008%20 Proceedings%20Documents/ETS%202008/Levine%20Enabling%20 Affordable%20Ships%20Through%20Physical%20Modular%20Open%20 Systems.pdf

DeVries, R., K. T. Tompkins, and J. Vasilakos, "Total Ship Open Systems Architecture," *ASNE Naval Engineers Journal*, Vol. 112, No. 4, July 2000.

Drewry, John T., and Otto P. Jons, "Modularity: Maximizing the Return on the Navy's Investment," *Naval Engineers Journal*, Vol. 87, No. 2, April 1975, pp. 198–214.

Duren, B. G., and J. R. Pollard, *Total Ship System Engineering Vision and Foundations,* Dahlgren, Va.: Naval Surface Warfare Center Dahlgren Division, 1995.

Estabrook, A., R. MacDougall, and R. Ludwig, *Unmanned Air Vehicle Impact on CVX Design*, San Diego, Calif.: Space and Naval Warfare Systems Center, Technical Document 3042, September 1998.

Eckstein, Megan, "Admirals: Modular, Multimission Ships a Must in Future Surface Fleet," *Inside the Navy*, January 25, 2013.

Eckstein, Megan, "Shannon: Navy Ready to Upgrade Cruisers if Congress Provides Funding," *Inside the Navy*, May 12, 2012.

Federation of American Scientists, *Surface Officer Warfare School Documents: 64P7-101 GENERAL DESCRIPTION FFG 7*, undated a. As of June 18, 2014: http://www.fas.org/man/dod-101/navy/docs/swos/eng/64P7-101.html

Federation of American Scientists, *Surface Officer Warfare School Documents: PS8-101 GENERAL DESCRIPTION DD-963, DDG-993, and CG-47*, undated b. As of June 18, 2014: http://www.fas.org/man/dod-101/navy/docs/swos/eng/PS8-101.html

Fein, Geoff, "USN Issues Open System Strategy to Foster Component Use Across Multiple Platforms," *International Defence Review*, January 11, 2013.

Fireman, Howard, Marianne Nutting, Tom Rivers, Gary Carlile, and Kendall King, "LPD 17 on the Shipbuilding Frontier: Integrated Product and Process Development," *35th Annual Technical Symposium*, Association of Scientists and Engineers, April 17, 1998.

French, Daniel W., *Analysis of Unmanned Undersea Vehicle (UUV) Architectures and an Assessment of UUV Integration into Undersea Applications*, thesis, Monterey, Calif.: Naval Postgraduate School, September 2010.

Flyvefisken-class STANFLEX 300 ships, GlobalSecurity.org, undated. As of April 2012: http://www.globalsecurity.org/military/world/europe/hdms-flyvefisken.htm

Frank, Matthew V., and Richard Helmick, *21st Century HVAC System for Future Naval Surface Combatants—Concept Development Report*, Carderock, Md.: Naval Surface Warfare Center Carderock Division, 2007.

————, "21st Century Heating Ventilation and Air Conditioning (HVAC) System for Future Surface Combatants," *ASNE Proceedings 2008*, Alexandria, Va.: American Society of Naval Engineers, 2008.

Garver, S. N., and J. Edyvane, "Ship Modularity Cost-Reduction Models," *ASNE Proceedings 2010,* Alexandria, Va.: American Society of Naval Engineers, 2010.

Garver, S., R. Marcantonio, and P. Sims, "Modular Adaptable Ship (MAS) Total Ship Design Guide for Surface Combatants," Naval Sea Systems Command, SER 9/05T, February 7, 2011.

Galye, Wayne, *Analysis of Operational Manning Requirements and Deployment Procedures for Unmanned Surface Vehicles Aboard US Navy Ships*, thesis, Monterey, Calif.: Naval Postgraduate School, March 2006.

Geer, Harlan, and Christopher Bolkcom, *Unmanned Aerial Vehicles: Background and Issues for Congress*, Washington, D.C.: Congressional Research Service, RL31872, November 21, 2005.

General Accounting Office, *Status of the Navy's Aegis Weapon System and DDG-47 Shipbuilding Program*, Washington, D.C., C-PSAD-80-18, February 28, 1980.

Gertler, Jeremiah, *U.S. Unmanned Aerial Systems*, Washington, D.C.: Congressional Research Service, R42136, January 3, 2012. As of February 22, 2014:
http://www.fas.org/sgp/crs/natsec/R42136.pdf

Gibbs & Cox, Inc., *Draft Engineering Change Proposal for MGS Installation on DDG 51 Flight IIA Class Ships*, January 24, 1997.

Grant, Benjamin P., *Density as a Cost Driver in Naval Submarine Design and Procurement*, thesis, Monterey, Calif.: Naval Postgraduate School, June 2008.

Greenert, Jonathan W., U.S. Navy, "Payloads over Platforms: Charting a New Course," *Proceedings*, Vol. 138/7/1,313, July 2012. As of June 18, 2014:
http://www.usni.org/magazines/proceedings/2012-07/
payloads-over-platforms-charting-new-course

Hone, Thomas C., Norman Friedman, and Mark D. Mandeles, *American and British Aircraft Carrier Development, 1919–1941*, Annapolis, M.: Naval Institute Press, 1999.

Hoorens, Stijn, Jeremy Ghez, Benoit Guerin, Daniel Schweppenstedde, Tess Hellgren, Veronika Horvath, Marlon Graf, Barbara Janta, Samuel Drabble, and Svitlana Kobzar, *Europe's Societal Challenges: An Analysis of Global Societal Trends to 2030 and Their Impact on the EU,* Santa Monica, Calif.: RAND Corporation, RR-479-EC, 2013. As of June 18, 2014:
http://www.rand.org/pubs/research_reports/RR479.html

Jaishankar, C., "UAV Squadron to Come Up to Uchipuli," *The Hindu*, April 8, 2012. As of July 3, 2012:
http://www.thehindu.com/news/cities/chennai/article3291364.ece

Jolliff, J. V., "Modular Ship Design Concepts," *Naval Engineers Journal*, Vol. 86, No. 5, October 1974, pp. 11–30.

Jones, R. V., *The Wizard War: British Scientific Intelligence 1939–1945*, London: Hamish Hamilton, 1978.

Karvar, Patrick, Shawna Garver, Ray Marcantonio, and Philip Sims, *Modular Adaptable Ship (MAS) Total Ship Design Guide for Surface Combatants*, Washington, D.C.: Naval Sea Systems Command, February 2011.

Keane Jr., R. G., "Reducing Total Ownership Cost: Designing Inside-Out of the Hull," *ASNE Day 2011 Proceedings*, Alexandria, Va.: American Society of Naval Engineers, 2011.

Lawson, Charles E., "SEAMOD—A New Way to Design, Construct, Modernize, and Convert U.S. Navy Combatant Ships," *14th Annual Technical Symposium*, Association of Scientists and Engineers, 1977.

———, "SEAMOD—A New Way to Design, Construct, and Modernize Navy Combatant Ships," *Naval Engineers Journal*, Vol. 90, No. 1, 1978, pp. 3–108.

Levine, Andrew J., William H. Mish Jr., and Timothy M. Lynch, "Application of Physical Open Systems to Meet Technological Requirements and Capabilities—A Modular Reconfigurable Space," *ASNE Day 2008 Proceedings*, Alexandria, Va.: American Society of Naval Engineers, 2008. As of February 22, 2014: https://www.navalengineers.org/SiteCollectionDocuments/2008%20 Proceedings%20Documents/ASNE%20Day%202008/paper30.pdf

Lundquist, E. H., "USS *Bunker Hill* Emerging from Transformation as a 'Sharper Sword': Cruise Modernization Extends Combatant Service Life, Enhances Capabilities," *Naval Engineers Journal*, Vol. 120, No. 3, 2008.

Marcantonio, Raymond T., E. Gregory Sanford, David S. Tillman, and Andrew S. Levine, "Addressing the Design Challenges of Open System Architecture Systems on U.S. Navy Ships—Building Out of the Box," MAST 2007 Conference, 2007.

Marine Board, Commission on Engineering and Technical Systems, *Toward More Productive Naval* Shipbuilding, Washington, D.C.: National Academy Press, 1984.

McCoy, Timothy J., *Electric Ships Office Overview*, briefing presented at Surface Navy Association National Symposium, Arlington, Va., January 12, 2011.

Naval Surface Warfare Center Dahlgren Division, *Open Architecture (OA) Computing Environment Design Guidance*, Version 1.0, August 23, 2004.

Naval Vessel Register, Web page, undated. As of June 2012: http://www.nvr.navy.mil/nvrships/

Office of the Chief of Naval Operations, *Report to Congress on Annual Long-Range Plan for Construction of Naval Vessels for FY 2011*, Washington, D.C.: Department of the Navy, OPNAV N8F, February 2010.

Office of the Under Secretary of Defense (Acquisitions, Technology, and Logistics), Property and Equipment Policy Office and Office of the Under Secretary of Defense (Comptroller), Accounting and Finance Policy Office, *Military Equipment Useful Life Study—Phase II, Final Report*, Washington, D.C., May 30, 2008.

Open Systems Joint Task Force, *Program Manager's Guide: A Modular Open Systems Approach (MOSA) to Acquisition*, U.S. Department of Defense, Version 2.0, September 2004.

O'Rourke, Ronald, *Navy DDG-1000 Destroyer Program: Background, Oversight Issues, and Options for Congress*, Washington, D.C.: Congressional Research Service, RL32109, June 5, 2008.

O'Rourke, Ronald, *Navy Shipboard Lasers for Surface, Air, and Missile Defense: Background and Issues for Congress*, Washington, D.C.: Congressional Research Service, R41526, April 8, 2011. As of February 22, 2014: http://www.fas.org/sgp/crs/weapons/R41526.pdf

Rocky Mountain Institute, *Energy Efficiency Survey Aboard USS* Princeton *CG-59*, June 30, 2001.

Savitz, Scott, Irv Blickstein, Peter Buryk, Robert W. Button, Paul DeLuca, James Dryden, Jason Mastbaum, Jan Osburg, Philip Padilla, Amy Potter, Carter C. Price, Lloyd Thrall, Susan K. Woodward, Roland J. Yardley, and John M. Yurchak, *U.S. Navy Employment Options for Unmanned Surface Vehicles (USVs)*, Santa Monica, Calif.: RAND Corporation, RR-384-NAVY, 2013. As of June 24, 2014: http://www.rand.org/pubs/research_reports/RR384.html

"ScanEagle, United States of America," *Naval-Technology.com*, undated. As of February 5, 2014: http://www.naval-technology.com/projects/scaneagle-uav/

Schank, John F., Hans Pung, Gordon T. Lee, Mark V. Arena, and John Birkler, *Outsourcing and Outfitting Practices: Implications for the Ministry of Defence Shipbuilding Programmes*, Santa Monica, Calif.: RAND Corporation, MG-198-MOD, 2005. As of June 18, 2014: http://www.rand.org/pubs/monographs/MG198.html

Scott, Richard, "Frigate Deploys with Four Fire Scout UAVs," *Jane's Navy International online*, posted July 2, 2012.

Tilghman, Andrew, "New Rating Considered for UAV Operators," *Navy Times*, November 2, 2008. As of June 25, 2012: http://www.navytimes.com/news/2008/11/navy_uav2_110208w/

Treverton, Gregory F., *Making Policy in the Shadow of the Future*, Santa Monica, Calif.: RAND Corporation, OP-298-RC, 2010. As of June 18, 2014: http://www.rand.org/pubs/occasional_papers/OP298.html

Trumble, John C., John J. Dougherty, Laurent Deschamps, Richard Ewing, Charles R. Greenwell, and Thomas Lamb, *Product Oriented Design and Construction (PODAC) Cost Model—An Update*, paper presented at the 1999 Ship Production Symposium, July 29–30, 1999.

Truver, Scott C., "Navy Develops Product Oriented Design and Construction Cost Model," *Program Manager*, Vol. 30, No. 1, January/February 2001.

UAV GROUND CONTROL STATION (GCS) Basis of Issue Plan, undated. As of February 22, 2014:
http://www.fas.org/irp/program/collect/docs/bnM066AE.htm

U.S. Navy, "U.S. Navy Fact Sheet: MK 45—5-inch-gun 54/62 Caliber Guns," *Navy.mil*, October 19, 2012. As of December 2012:
http://www.navy.mil/navydata/fact_print.asp?cid=2100&tid=575&ct=2&page=1

U.S. Navy, "U.S. Navy Fact Sheet: Attack Submarines—SSN," *Navy.mil*, December 6, 2013. As of March 20, 2014:
http://www.navy.mil/navydata/fact_print.asp?cid=4100&tid=100&ct=4&page=1

Vasilakos, J., R. Marcantonio, and S. Garver, "A Guide for the Design of Modular Zones on US Navy Surface Combatants," Naval Sea Systems Command, SER-4/05T, January 25, 2011.

Webster, Wade A., and William D. Tootle, "Is It SEAMOD Time?" *28th Annual Technical Symposium*, Association of Scientists and Engineers, 1991.

Whitmer, Lynden D., *Naval UAV Programs: Sea Based UAV's*, Dahlgren, Va.: Naval Surface Warfare Center Dahlgren Division, February 26, 2002.

Work, Robert O., *Naval Transformation and the Littoral Combat Ship*, Washington, D.C.: Center for Strategic and Budgetary Assessments, 2004.